Quality Assurance of Chemical Measurements

JOHN KEENAN TAYLOR

Library of Congress Cataloging-in-Publication Data

Taylor, John K. (John Keenan), 1912–
Quality assurance of chemical measurements.

Bibliography: p.
Includes index.
1. Chemical laboratories—Quality control.
I. Title.
QD51.T38 1987 542′.3 86-27335
ISBN 0-87371-097-5

3rd Printing 1988
2nd Printing 1987

LEWIS PUBLISHERS, INC.
121 South Main Street, P.O. Drawer 519, Chelsea, Michigan 48118

PRINTED IN THE UNITED STATES OF AMERICA

Preface

Chemical measurement data are often the basis for critical decisions on vital matters, ranging from the health of individuals, to the protection of the environment, to the production of safe, reliable, and useful products and services. Obviously, the data used for such purposes must be reliable, and there must be unequivocal evidence to prove it. The philosophy and procedures by which this is achieved and demonstrated are called quality assurance.

This book discusses the basic concepts of quality assurance as presented in a short course given by the author many times since 1979. It is based on the premise that a good understanding of the basic principles of chemical measurement is a necessary prerequisite to achieving accurate data and to designing and implementing a credible quality assurance program.

Credible data must be supported by probabilistic statements of the confidence that can be placed in it. This requires a systematic approach to measurement and to the attainment of statistical control of the measurement process by quality control procedures. The precision and accuracy of the data so produced can be evaluated by quality assessment techniques.

Not only the philosophy, but also practical approaches to quality assurance are discussed, and the statistical techniques vital to its various aspects are reviewed.

The book is written to provide guidance for the development of a credible quality assurance program and also for its implementation. The chapters are presented in a logical progression starting with the concept of quality assurance, the principles of good measurement, the principles of quality assurance, and the evaluation of measurement quality. Guidance is provided for the development of quality assurance programs and for the improvement of existing ones. Each chapter has a degree of independence so that it may be consulted, in isolation from the others, when its subject matter is of interest in a particular situation. An extensive appendix containing definitions of quality assurance terminology, a collection of useful statistical tables, outlines useful for guidance in the preparation of quality assurance program documents, study aids for those using the book for self instruction or as a text, and an extensive collection of abstracts of important quality assurance publications are also included. The abstracts also serve as the list for references cited in the text.

If the book achieves the above-mentioned objectives, it will have provided a valuable service to the producers and users of chemical measurement data.

However, it can serve another purpose as well—namely, in training analytical chemists. Most of modern analytical education is devoted to the theory and principles of measurement techniques, and such knowledge is a necessary foundation for reliable measurements. The present book supplements such information and provides guidance in the practice of the profession of analytical chemistry.

While the text focusses principally on chemical measurement, the author is confident that the material presented is widely applicable to the physical and biological sciences as well. The basic principles of measurement are the same, no matter what the subject area of application. In fact, he has developed quality assurance programs for high accuracy physical measurement and calibration programs that do not differ in kind, but only in detail, from the chemical programs described in this book. The biological scientist should also find the quality assurance principles enunciated useful in measurement programs.

Acknowledgments

The author is grateful to all of the excellent analytical chemists with whom he has been privileged to associate in various ways over many years, who have contributed to his education and have helped to mould his philosophy of good measurement. Unfortunately, most of them must go unnamed because of space limitations. The subject matter presented in the book is the result of the author's gleanings from many sources, including his long association with like-minded colleagues at the National Bureau of Standards (NBS), who have had measurement accuracy as their personal as well as institutional goals. He grew up under the tutelage of the late Dr. G.E.F. Lundell, one of the early deans of accurate chemical measurements. His concepts of the critical importance of sampling and the analytical sample were sharpened and increased by a year's sabbatical association with Professor Byron G. Kratochvil of the University of Alberta. His long personal involvement as a bench chemist, and later as a supervisor responsible for reference material analysis and certification, have provided practical insight to constructive planning and to the avoidance of measurement pitfalls.

The author was very fortunate to have worked under the late Leroy W. Tilton, who was a master metrologist and a pioneer in the application of statistics to experimental measurements. More recently, he has benefited from association with and guidance from the NBS Statistical Engineering Division, and especially from Mrs. Mary G. Natrella. His recent position as Coordinator for Quality Assurance of the NBS Center for Analytical Chemistry has provided the opportunity to put all of this past experience together in a systematic fashion as a quality assurance concept, which provides the substance for this book.

John K. Taylor is an analytical chemist of many years of varied experience. All of his professional life has been spent at the National Bureau of Standards, from which he recently retired after 57 years of service.

Dr. Taylor received his BS degree from George Washington University and MS and PhD degrees from the University of Maryland. At the National Bureau of Standards, he served first as a research chemist, and then managed research and development programs in general analytical chemistry, electrochemical analysis, microchemical analysis, and air, water, and particulate analysis. The development of reliable sampling plans was another area of interest. The development and certification of reference materials was a major activity throughout his career at NBS. For the past seven years, he coordinated the NBS Center for Analytical Chemistry's program in quality assurance, and conducted research activities to develop advanced concepts to improve and assure measurement reliability. He provided advisory services to other government agencies as part of his official duties.

As a part of his recent activities, he has developed a comprehensive short course—Quality Assurance of Chemical Measurements—that has been presented some 75 times to a cumulative audience of over 3200 persons.

Dr. Taylor has edited three books, and has written over 200 research papers in analytical chemistry, some 20 of which have dealt with various aspects of the quality assurance of measurements.

Dr. Taylor is the recipient of several awards for his accomplishments in analytical chemistry, including the Department of Commerce Silver and Gold Medal Awards. He is a past chairman of the Washington Academy of Sciences and the ACS Analytical Chemistry Division, and is currently chairman of ASTM Committee D 22 on Sampling and Analysis of Atmospheres.

Table of Contents

List of Figures

List of Tables

Quality Assurance of Chemical Measurements

CHAPTER 1

Introduction to Quality Assurance

The demand for measurement data is ever increasing. Decisions need to be made on such questions as the suitability of a material for an intended purpose, the quality of the environment, and the health of individuals. The vast majority of such measurements entail chemical analyses in which some chemical property of the material, object, or system of interest is measured.

Dramatic changes have taken place in chemical analysis during the past several decades. Most measurements are now being made using complex instruments in sophisticated processes. Moreover, measurements made by several analysts and/or laboratories often need to be interrelated for use in a decision process. Monitoring programs of regional, national, or international scope may be involved. All of these developments involve stringent requirements for the reliability and compatibility of the data.

The quality of data is ordinarily evaluated on the basis of its uncertainty when compared with end-use requirements. If data have consistency and the uncertainty is small when compared to the requirements, they are considered to be of adequate quality. When excessively variable or the level of uncertainty exceeds the needs, the data may be said to be of low or inadequate quality. The evaluation of data quality is thus a relative determination. What is high quality in one situation could be unacceptable in another.

The quality of chemical data may be judged on the basis of two aspects: the accuracy of identification of the parameter measured and the numerical accuracy. The qualitative identification must be beyond reasonable doubt. The analyst must state precisely what was measured and be able to prove it. While empirical measurements defined by a specific apparatus or methodology may be useful in some situations, they must be specified as such and used with due consideration of their limitations.

Quantitative measurements are always estimates of the value of the measure and involve some level of uncertainty. The measurements must be made so that the limits of uncertainty can be assigned within a stated probability. Without such an assignment, no logical use can be made of the data. To achieve this, measurements must be made in such a way to provide statistical predictability. The experience of metrologists has demonstrated that this is achieved best by a well-designed and consistently implemented quality assurance program.

Quality assurance consists of two separate but related activities: quality control and quality assessment. Both must be operational and coordinated. The following definitions are offered [132].*

> Quality assurance: A system of activities whose purpose is to provide to the producer or user of a product or a service the assurance that it meets defined standards of quality with a stated level of confidence.
>
> Quality control: The overall system of activities whose purpose is to control the quality of a product or service so that it meets the needs of users. The aim is to provide quality that is satisfactory, adequate, dependable, and economic.
>
> Quality assessment: The overall system of activities whose purpose is to provide assurance that the overall quality control job is being done effectively. It involves a continuing evaluation of the products produced and of the performance of the production system.

The objective of quality control is to fine tune and maintain a measurement process in a desired state of stability and reproducibility. In such, the process can be considered to have a high degree of similarity to an industrial production process, with the ability to produce objects (in this case data) with a high degree of reproducibility. Once this is established, the precision may be defined and any biases can usually be identified and eliminated or compensated, so that requisite data quality is achieved. Quality assessment provides the evidence that this has been accomplished.

The general approach to quality assurance of the measurement process parallels that for the industrial production process and borrows concepts and techniques proven trustworthy for the latter, to the extent possible. In industry, the procedure may be called statistical product control. In metrology, the procedure may be called statistical measurement control.

The quality assurance procedure followed in a typical production process is illustrated in Figure 1.1. Experience dictates the kind and degree of control that must be achieved and maintained to obtain products of desired quality. Random samples of the output are tested or inspected with respect to specifications in order to release or accept the production lot, to reject it, or to take corrective actions in the process.

The quality assurance aspects of a typical measurement process are illustrated in Figure 1.2. Quality control procedures are used to tune and maintain the system in a state of statistical control in which it may be considered as

*The numbers in brackets here and throughout the book are citations to publications in the Bibliography contained in Appendix E.

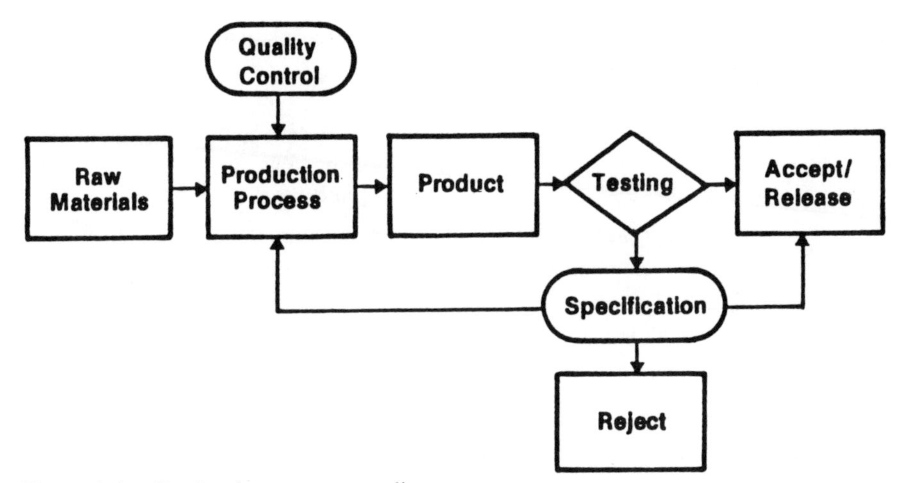

Figure 1.1. Production process quality assurance.

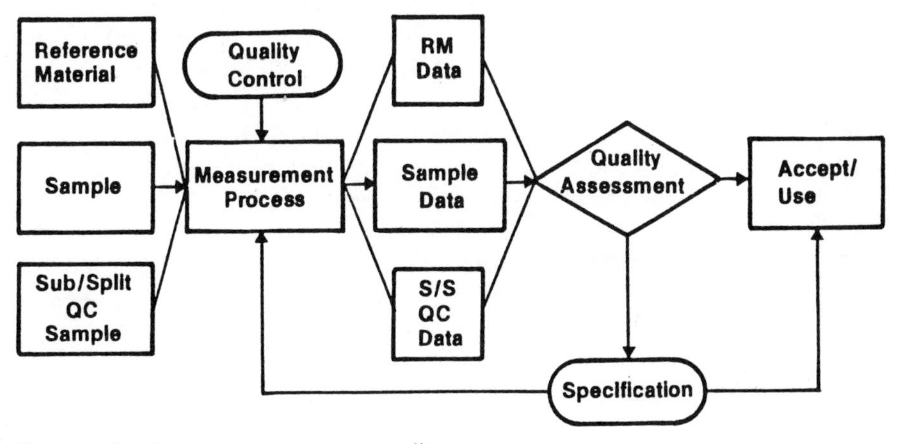

Figure 1.2. Measurement process quality assurance.

capable of generating an infinite number of measurements on any material, of which the data of the moment are a typical sample. Quality assessment procedures are then used to evaluate the quality of the data that is produced. Unfortunately, it is generally impossible, intrinsically, to evaluate fully the quality of the data on unknown test materials. However, if the measurement system is maintained in a state of statistical control, known samples such as reference materials (RM), which simulate the test samples, may be measured concurrently and the results compared with the reference values. Such a comparison can be used to evaluate the performance of the measurement system and permit inferences to be made on the quality of the data for test samples.

It should now be evident that statistics is an integral part of quality assurance. In fact, what has really been discussed here is statistical assurance of quality. It may be looked at in the following way:

Statistical	Use of data from a stable measurement system
Assurance	to provide probabilistic confidence of achievement
Quality	of a desired level of acceptability

By common usage, the term has been shortened to quality assurance.

This discussion concerns the quality of measurement data, and indeed the quality of individual measured values. When considered collectively as a body of numerical information, additional quality descriptors may be useful. Thus the Environmental Protection Agency proposes [125]:

Completeness—A measure of the amount of data obtained from a measurement process compared to the amount that was expected to be obtained under the conditions of measurement.

Representativeness—The degree to which the data accurately and precisely represent a characteristic of a population parameter, variation of a property, a process characteristic, or an operational condition.

Comparability—The confidence with which one data set can be compared to another.

Although they are difficult to quantify, such characteristics need to be considered when evaluating the usefulness of a set of data for scientific interpretation.

In addition to meeting the requirements stated above, data must be both technically sound and legally defensible. That is to say, it must be defensible against any reasonable adversarial inquiry or action, legal or otherwise. There is a difference in these requirements and both need to be met. The first is necessary but not sufficient. It includes all those things that a careful competent analytical chemist should do. The second includes the proof that sound work was done. Good quality assurance practices are central to achieving both goals. They stress documentation which takes the burden off memory and can remove any suspicion or shadow of doubt of details of what was done. Good quality assurance practices cannot give credence to poor technical work. In fact, they can help to identify poor quality when it exists. But, lack of documentation can make defense deficient if not impossible. While legal defense in its broadest connotation is often the issue, technical reliability is always at stake. Quality assurance practices can promote this as well. In fact, measurements should always be made, believing that there is someone somewhere who may have some questions about what was done. When questions are anticipated, measurement plans often can be adjusted to answer them and to document what is done with little or no extra effort.

Before ending this introductory discussion, we should remember that there are two aspects of quality. First, there is design quality which is based on a plan to meet an envisioned need. This is essential if anything useful is to result from any production process, whether it provides products, services, or data. The second is conformance quality and is concerned with the achievement of design quality. The first does not guarantee the second, but the second is hardly achievable without the first.

Production of quality data does not occur automatically and is not guaranteed by following any one of several established procedures. It is enhanced by:

- understanding the measurement process
- understanding what needs to be measured
- understanding what needs to be done to obtain reliable measurements

Quality can be achieved by dedicated effort, by careful attention to detail, and by practicing recognized principles of quality assurance. Such efforts must be accompanied by a desire for excellence and the creation of an atmosphere conducive to ensuring quality.

Quality assurance is more than a program; it is a philosophy, a way of life. As a program that is mechanically followed, quality assurance is doomed to failure. As a philosophy, there is a chance for success. When it is approached as both a program and a philosophy, the chances for producing high quality data are excellent.

CHAPTER 2

Precision, Bias, and Accuracy

Accuracy is concerned with correctness. If a measurement process produces the correct results, it is accurate, and the measured value is also accurate. Strictly speaking, only a counting process is exact and hence accurate. A measured value is an estimate and may be "close enough" to the correct value to be considered "errorless," and the process so producing it may be considered to be accurate.

When closely scrutinized, repetitive measurements will differ from one another, and the means of sets will differ, however to a lesser degree. The scatter of the values is a measure of the precision; the less the scatter, the higher the precision. In a stable measurement process, a large number of individual values will tend to converge toward a limiting mean, which may or may not be the true value. If not, the process is said to be biased.

The above discussion leads to the following definitions:

Accuracy – The degree of agreement of a measured value with the true or expected value of the quantity of concern.

Bias – A systematic error inherent in a method or caused by some artifact or idiosyncrasy of the measurement system. Temperature effects and extraction inefficiencies are examples of the first kind. Blanks, contamination, mechanical losses, and calibration errors are examples of the latter kinds. Bias may be both positive and negative, and several kinds can exist concurrently so that net bias is all that can be evaluated, except under special conditions.

Precision – The degree of mutual agreement characteristic of independent measurements as the result of repeated application of the process under specified conditions.

The concepts of accuracy, bias, and precision are illustrated in Figure 2.1 and Figure 2.2. All of the measurement processes of Figure 2.1 are unbiased, but differ in their accuracies due to their widely differing precisions. All of the processes of Figure 2.2 are inaccurate, based on considerations of bias. How-

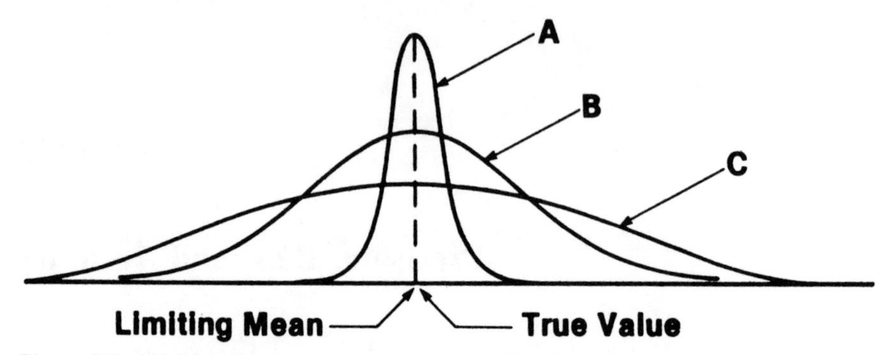

Figure 2.1. Unbiased measurement processes. The distributions of results from three unbiased processes are shown. The precision decreases in the order A>B>C. While the limiting means of all will approach the "true value," process C is relatively inaccurate (compared with A) due to its large imprecision.

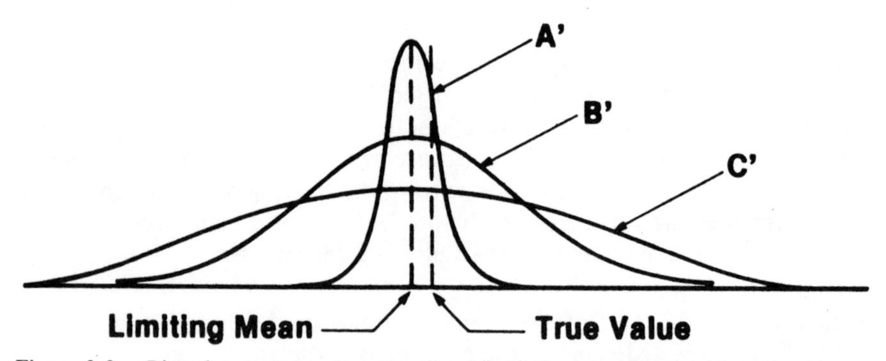

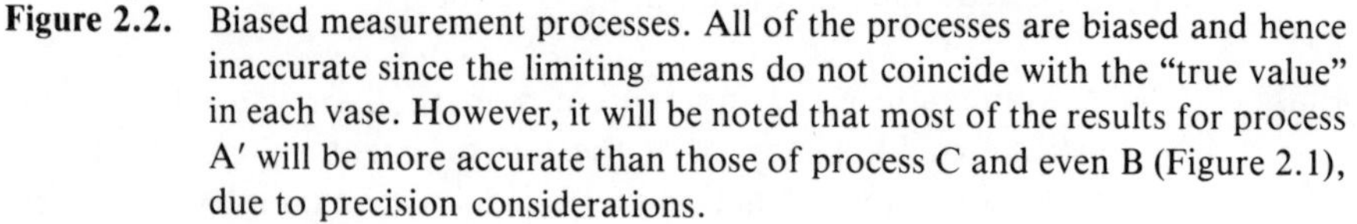

Figure 2.2. Biased measurement processes. All of the processes are biased and hence inaccurate since the limiting means do not coincide with the "true value" in each vase. However, it will be noted that most of the results for process A′ will be more accurate than those of process C and even B (Figure 2.1), due to precision considerations.

ever, it is evident that a biased process such as A is capable of producing data of higher accuracy than an unbiased process such as C, due to precision considerations.

AXIOMS

A measurement is . . .

Accurate when the value reported does not differ from the *true* value.

Biased when the error of the limiting mean is not zero; influenced by systematic error.

COROLLARIES

Error in reported values occurs as a result of bias and imprecision.

An *accurate method* is one *capable* of providing precise and unbiased results (within acceptable limits). *Accurate* and *precise* are relative terms.

In practice, we evaluate inaccuracy. Likewise, we evaluate imprecision, namely, the deviations of measurements.

The terms accurate, inaccurate, precise, and imprecise are relative to the end use of the data. A measurement process capable of reproducing the same value within 10% would be considered to be precise (even highly precise) in ultratrace analysis but very imprecise (in fact, useless) for major constituent analysis. Likewise, a bias of 5% (relative) would be considered insignificant in the former case and unacceptable in the latter case.

SOURCES OF ERROR

Measurement errors are of three types. Systematic errors are always of the same sign and magnitude and produce biases. They are constant no matter how many measurements are made. Random errors vary in sign and magnitude and are unpredictable. In other words, they occur by chance. Random errors average out and approach zero if enough measurements are made. Blunders are simply mistakes that occur on occasion and produce erroneous results. Measuring the wrong sample, errors of transcription or transposition of measured values, misreading a scale, and mechanical losses, are examples of blunders. They produce outlying results that may be recognized as such by statistical procedures, but they cannot be treated by statistics. Appropriate quality control procedures can minimize the occurrence of some kinds of blunders but may not eliminate carelessness which often is their principal cause.

In addition, there can be random components of systematic errors. For example, an instrument zero reading is adjusted to eliminate a systematic error but this adjustment has a certain amount of imprecision. Any single adjustment of a zero will produce a bias due to the inability to do this exactly. However, over many adjustments, the imprecision of a zero adjustment should cancel out. The uncertainty of this adjustment can be the imprecision of the setting process.

Experimental operations such as the above can be influenced by the biases of the operator. Thus one experimenter may favor high readings, and another may habitually set on the low side. The observations of each may be self consistent, but incompatible with each other. Experimenters making visual observations, such as the reading of scales, have been shown to have preferences for certain scale values.

ERRORS OF ROUNDING

While variability is a natural characteristic of measurement data, observers, or the instruments themselves may round off data at some point so that natural variability may be obscured. The observer may have a preconceived idea of the significance of measured values and round off to a value considered to be "meaningful." Also, sufficient care may not be exercised to push observations to their ultimate. Likewise, measuring instruments may be damped to the extend that essentially the same reading is obtained for each measurement. All of these practices tend to "edit" data, sometimes to the extent that its statistical character cannot be evaluated.

It is recommended that observational data should be taken with as many significant figures as possible and that rounding be deferred until its full statistical significance is evaluated, which ordinarily means after all calculations have been made.

INFLUENCE OF UNCERTAINTY ON DECISIONS

The analytical uncertainty must always be known when using data to make decisions. One example of the influence of analytical uncertainty on a decision process is illustrated in Figure 2.3. In this case, the question is simply does a measured value indicate that the analyte concentration exceeds some decision level, D. For a selected level of confidence, measured values less than A are clearly smaller than D, while values larger than B are clearly greater than D. A and B thus mark the levels of indecision. One cannot say, at the selected level of confidence, whether values within the limits A to B exceed or are less than the decision level. One of the prime objectives of quality assurance is to evaluate measurement uncertainty. When this is ascertained, the width of the band A to B can be known and one can decide whether this is large enough to influence adversely the decision process. Clearly, the width of the band must be known before any use can be made of the data for the purpose above.

The preceding discussion has assumed that the measurements involved were made by an unbiased measurement process. A biased process would underestimate or overestimate the concentration by a constant amount, depending upon the sign and magnitude of the bias, in addition to any random error of

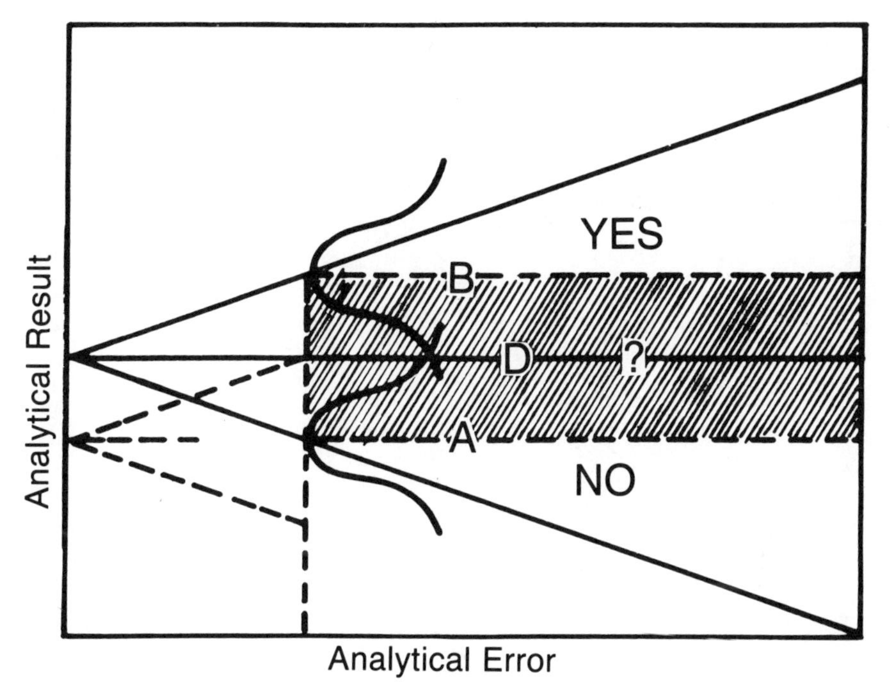

Figure 2.3. Influence of analytical uncertainty.

measurement. Adequate calibration should minimize the influence of bias on measurement decision.

If the band is too wide for practical purposes, it can be decreased by basing the decision on the average of a sufficient number of measurements, by improving the precision of the measurement process, or by using a more precise method of measurement, if available. Both availability of methodology and cost considerations will be involved when considering what approach to follow. If the decision maker is willing to take a bigger risk of making the wrong decision, the width of the band can be decreased.

The uncertainty considered in the foregoing example could arise from measurement and/or inhomogeneity of sample. These will be discussed more fully in Chapter 8. The statistical techniques useful for assigning uncertainties are discussed in Chapter 4.

CHAPTER 3

Statistical Control

It is the common experience of careful analysts that a number of independent measurements of the same quantity will produce a variety of results. In fact, variability is the hallmark of good measurement, and one should be suspicious if the results of individual measurements agree too closely. Even the means of several measurements will disagree although this disagreement should be less than that for the individual measurements.

On the other hand, if the results of repeated measurements disagree too much, one does not know which if any of them to believe. The output of a measurement process becomes believable as it attains a state described as "statistical control." It is said to have attained such a state when the means of a large number of individual values tend to approach a limiting value called the limiting mean. Moreover, the individual values will be distributed about the limiting mean in a stable distribution described by a stable standard deviation. The data output of a system in statistical control has statistical predictability, and various statistical calculations may be made to describe and evaluate it.

The term statistical control is somewhat of a misnomer. It does not mean that statistics controls the process but that there is statistical evidence that it is in control. Statistical control results from quality control of a measurement process. All critical variables are controlled to the extent necessary and possible, resulting in stability of the process and reproducibility of the data within defined limits. It has been stated [46]:

> ". . . Until a measurement operation has attained a state of statistical control, it cannot be regarded in any logical sense as measuring anything at all."

There is no proof for this statement, but its truthfulness is almost axiomatic. Measurement results can only be credible as some degree of consistency is attained.

While essentially mandatory, the attainment of statistical control may be difficult to prove. Continuous monitoring of a system to demonstrate its stability is the best if not the only approach. In fact, evidence of attainment is best based on lack of evidence of departure from statistical control. The use of appropriate control charts is the best way to document statistical control.

Statistical control is necessary to evaluate the precision of a process but says nothing about bias. However, it is a prerequisite for the evaluation of bias. It is useless to try to identify bias until statistical control has been attained and the variability of the process has been evaluated.

Statistical control does not indicate that a measurement process has been optimized, but only that it has been stabilized. Similar measurement processes may be operating in a state of statistical control in several laboratories, for example, but the respective performance parameters may differ significantly. The individual precisions may differ, and application-related biases may be present. However, the mere existence of such cannot be demonstrated unless each laboratory has attained statistical control. Only then is it possible to seek corrective actions that could enhance a greater compatibility of the various data outputs.

CHAPTER 4

Statistical Techniques

Every practicing analytical chemist must have a good understanding of the statistics of measurement. Statistical principles are applied when demonstrating statistical control, evaluating data, designing measurement and sampling plans, and making a wide variety of decisions in the use and application of measurements. Fortunately, modern calculators, computers, and data processing systems have built-in statistical packages that lessen the labor of statistical application, but they do not eliminate the requirement for understanding. In fact, the improper use of statistics and their inappropriate application can be as hazardous as using faulty data to make decisions.

This chapter is included to review the basic concepts of the statistics of measurements and to provide guidance in their application. Because of the small amount of data that ordinarily is available, statistical judgments should not be followed blindly. Whenever there is conflict between an intuitive and a statistically-based conclusion, the basis for both conclusions should be reviewed. Perhaps something has been overlooked in the reasoning process, or the statistical model may be faulty. Perhaps more data may be needed to strengthen the statistically-based decision. Reconciliation of any differences could lead to a better understanding of the measurement process and improvement of the present as well as future sets of data.

BASIC REQUIREMENTS

The basic requirements for applying statistics to measurement data are:

- the measurement system is stable
- the individual measurements are independent of one another
- the individual measurements are random representatives of the population of data that could be produced

This means that the measurement process must be in a state of statistical control.

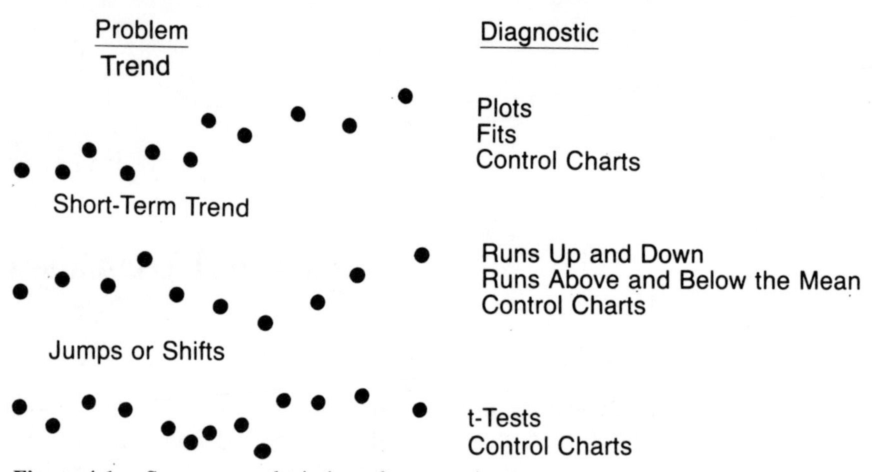

Figure 4.1. Some gross deviations from randomness.

The same three considerations apply to samples of materials that may be examined, namely:

- the population is stable
- the individual samples are independent of one another
- the individual samples are randomly selected from the population of interest

This means that the sampling process must be in a state of statistical control.

It is difficult, if not impossible, to confirm that these requirements have been met in any measurement or sampling operation. Rather, one has to look for evidence of nonconformance, and there are statistical tests to judge the significance of any suspected systematic deviations. If no significant deviations are found, one infers that the requirements are met. Because of the insensitivity of some statistical tests when the data base is small, other evidence such as physical knowledge about the measurement or sampling system may be needed to support judgment of attainment of statistical control.

Gross deviations from randomness can often be identified. Several types of these are illustrated in Figure 4.1, together with the statistical diagnostics that may be used to confirm the presence of a suspected problem. Control charts are an excellent way to demonstrate statistical control, and this is discussed in some detail in Chapter 14.

The question of independence of individual measurements is even harder to evaluate. A simple example of correlation is the use of differences of a series of data points. In such a case, a low reading at a preceding point could result in a low value for the preceding difference and a high value for the succeeding difference. Another common situation that can cause interdependence in a series of measurements occurs when all of them are based on the same calibration operation. If calibration is a significant source of variation in a measurement process, such a procedure could mask significant differences between data sets. As an example, if a measurement system is calibrated daily and ten measurements are made on each of two days, the measurements on each day are correlated with respect to calibration, and the statistical decision of the significance of the difference of the means would be based on nine degrees of freedom with respect to daily data but on only one degree of freedom with respect to between-day differences.

One should always be alert for correlations that would cause significant interdependence of measurement data and modify the procedure so that this does not constitute a problem. Statistical techniques that may be used to test for correlations are beyond the scope of this chapter, but they are discussed in a number of excellent books on statistical treatment of measurement data [100].

DISTRIBUTIONS

The term distribution refers to how the individual values spread around the mean. Ordinarily, data generated during a measurement process show the following properties [81]:

- the results spread symmetrically about a central value
- small deviations from the central value occur more frequently than large deviations
- the frequency distributions of a large body of data approximate a bell-shaped curve
- the means of even small sets of data show the above properties better than the individual values

If a large body of data is available, it is relatively easy to determine the nature of the distribution. Nelson [104] discusses frequency plots and other graphical ways to do this. For small numbers of data, variability makes judgment difficult. By chance, a distribution may be skewed, or its curvature (kurtosis) may be less or greater than normal. Several ways to quantitatively express the above situations are discussed elsewhere [7].

For most measurement processes in statistical control, the data may be assumed to be normally distributed for all practical purposes. However, it should be considered that deviations from normality can occur for such reasons as:

- presence of outliers that could falsely indicate skewness
- occasional shifts in operational characteristics that could skew the distribution (see Figure 14.5)
- undetected cycles that could cause apparent broadening of the distribution
- instability resulting in "up and down" fluctuations that would broaden the distribution

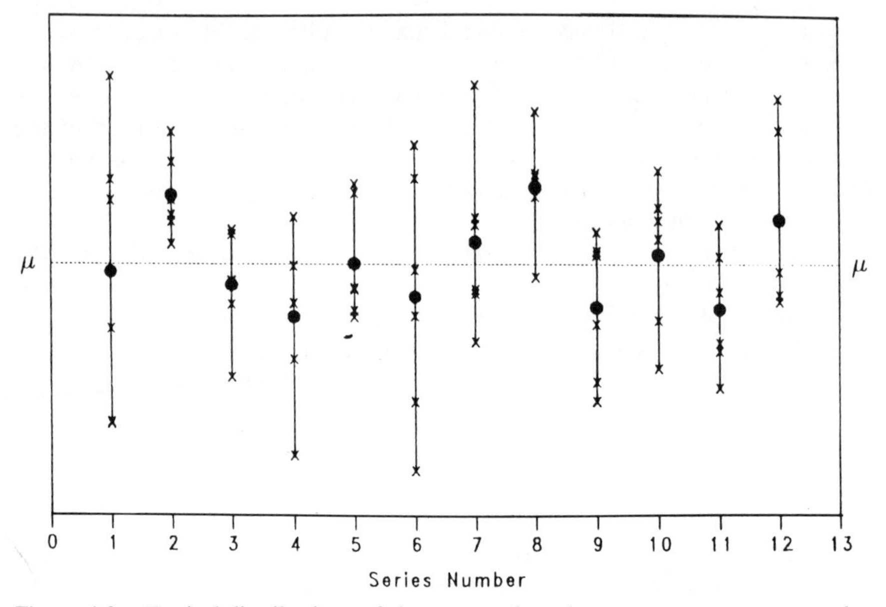

Figure 4.2. Typical distributions of the means of random measurements or samples.

When a control chart is maintained, and as the number of points increases, the correctness of an assumed normal distribution can be inferred from a symmetrical distribution of points about the central line, especially in the case of an X chart.

STATISTICS OF MEASUREMENTS

The following sections contain brief descriptions with examples of statistical calculations related to some of the questions that arise when evaluating chemical measurement data. The reader is referred to the many excellent books that are available which discuss these matters in more detail as well as the basis for the relationships used. General information on precision measurement is contained in Ku [82]; and NBS Handbook 91 [100] is especially recommended for a detailed discussion of statistical concepts. The latter contains many numerical examples as well as extensive tables, from which most of those included in Appendix C were taken.

The results of repetitive measurements are usually considered to be normally distributed and representable by a bell-shaped curve. If a series of n measurements were made many times, one would expect to obtain data sets represented by distributions such as those in Figure 4.2. The means of each set will differ from each other. If the sample standard deviation, s, is calculated for each set of such measurements, a different result would be expected each time. Fig-

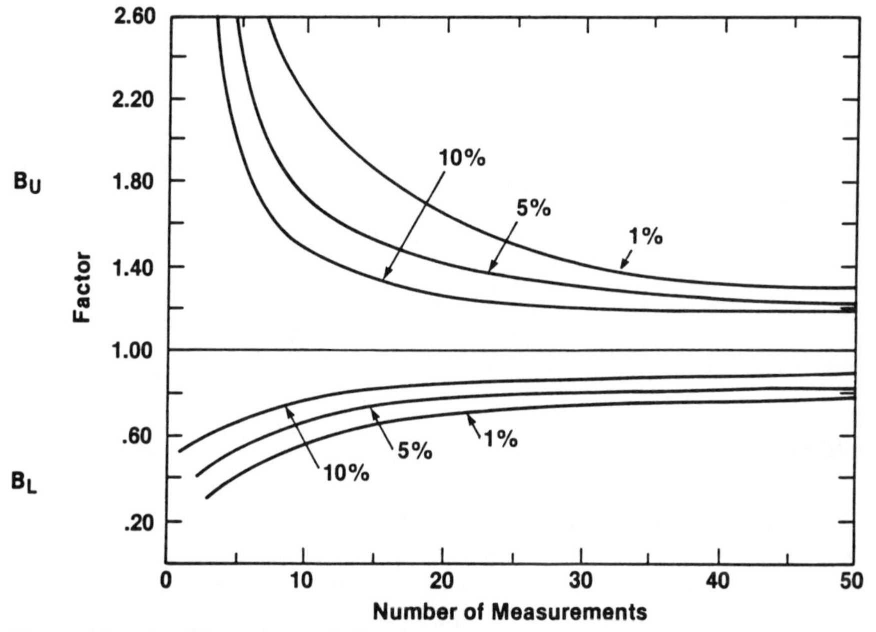

Figure 4.3. Confidence intervals for sigma.

ure 4.3 illustrates the expected variability of estimates of standard deviations made on various occasions as a function of the number of measurements involved. The factor plotted, when multiplied by the estimate of the standard deviation, gives the interval that is expected to include the population standard deviation for a given percentage of occasions. The labels, e.g., 10%, indicate the percentage of time that such an interval would not be expected to include σ. The factors used to construct the figure are found in Table C.4.

When several series of measurements are made, both the means and the standard deviations will vary from measurement to measurement, as illustrated in Figure 4.4. It will be noted that as n increases from 4 to 1000, the variation of the means decreases but never disappears.

On considering the above, it should be obvious that even the best of measurements will differ from each other, whether made by the same or different laboratories or scientists. The serious analytical chemist often needs to answer questions concerning the confidence that can be placed in measurement data and the significance of apparent differences resulting from measurements. The various equations given in this chapter take into account both the expected variability within populations and the uncertainties in the estimates of the population parameters that must be considered when answering such questions.

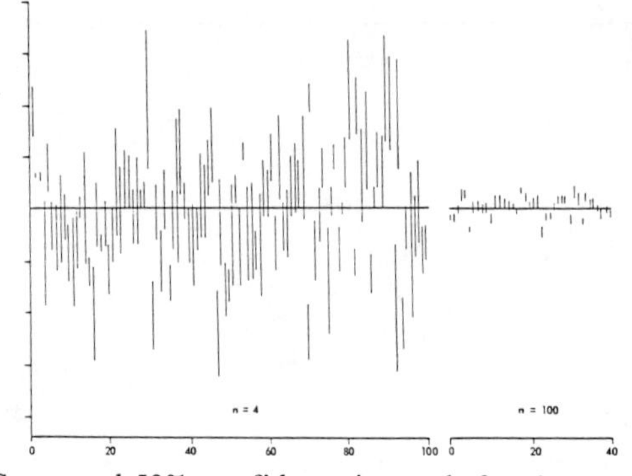

Figure 4.4. Computed 50% confidence intervals for the population mean, m.

ESTIMATION OF STANDARD DEVIATION

The basic parameters that characterize a population (universe) of samples or measurements on a given sample are the mean, μ, and the standard deviation, σ. Unless the entire population is examined, μ and σ cannot be known but are estimated from a sample(s) randomly selected (assumed) from it. The result is a sample mean, $\bar{X}$, and an estimate of the standard deviation, s, which must be used if such things as confidence intervals, population characteristics, tolerance intervals, comparisons of precision, and the significance of apparent discrepancies in measured values are to be evaluated. The variance, which is simply the standard deviation squared (e.g., σ^2, s^2) is used in some statistical calculations later in this chapter.

The standard deviation and its estimates represent the dispersion of the members of a population (the measured values) around the mean. The population (or sample of the population) that it represents must always be clearly identified; otherwise one cannot know to what it refers. In the case of measurement, it is helpful to consider a specific methodology, in a state of statistical control, as a generator of a population of values of which the data set at any time is a sample. If anything about the measurement process is changed on various occasions, the population produced may differ, as well as the standard deviation and mean. Such possibilities must be considered when statistically analyzing data sets. In the case of samples of materials, the parent material must be clearly identified when the dispersion of individual samples is discussed.

The standard deviation and its estimates have the same units as that of the population. Often, it is convenient to use dimensionless quantities to describe dispersions, namely the coefficient of variation (CV) and the relative standard deviation (RSD) defined as follows:

$$CV = \sigma/\text{mean or } s/\text{mean}$$

$$RSD = CV \times 100 \text{ (percent)}$$

A very important piece of information concerning standard deviation estimates is the number of degrees of freedom (symbol df or ν) upon which they are based. Without going into details, they signify the number of independent estimates that could be obtained from a specific set of data. In general, for a simple data set of n independent values,

$$df = n - 1$$

It will be noted that the value for df is required when determining the statistical quantities (e.g., t, k, and F) to be used in various calculations in the following sections.

Several ways by which the standard deviation may be estimated are given in the following sections.

Estimation of Standard Deviation from Replicate Measurements

The standard deviation is frequently estimated by making a number of replicate measurements of a given sample. The equations used to calculate the mean and to estimate the standard deviation are:

$$X_1, X_2, X_3, \ldots\ldots\ldots\ldots X_n$$

$$\bar{X} = \frac{X_1 + X_2 + X_3 + \ldots + X_n}{n}$$

$$s = \sqrt{\frac{\Sigma (X_i - \bar{X})^2}{n - 1}}$$

where s is estimated with $\nu = n - 1$ degrees of freedom.

Example: Series of Measurements

X	$(X_i - \bar{X})$	$(X_i - \bar{X})^2$
15.2	0.143	0.0204
14.7	-0.357	0.1257
15.1	0.043	0.0018
15.0	-0.057	0.0033
15.3	0.243	0.0590
15.2	0.143	0.0204
14.9	-0.157	0.0247
$\bar{X}$ = 15.057		Σ = 0.2572
		n = 7

$$s = \sqrt{\frac{0.2572}{6}} = 0.207$$

$$\nu = 6$$

Estimation of Standard Deviation from Duplicate Measurements

The standard deviation may be estimated from the differences of several sets of duplicate measurements. These may be duplicate measurements of the same sample on several occasions, or they may be duplicate measurements of different samples. In the latter case, the various samples included in the calculation must be similar, and the analyst must be sure that the precision of the measurement process is essentially the same for all samples included in the calculation. The equation used to make the estimation is:

$$s = \sqrt{\frac{\Sigma d^2}{2k}}$$

where k = number of sets of duplicate measurements
d = difference of duplicate measurements
ν = k degrees of freedom

Example A: Duplicates, Same Material

X_f	X_s	$\lvert d \rvert$	d^2
14.7	15.0	0.3	0.09
15.1	14.9	0.2	0.04
15.0	15.1	0.1	0.01
14.9	14.9	0.0	0.0
15.3	14.8	0.5	0.25
14.9	15.1	0.2	0.04
14.9	15.0	0.1	0.01
Σ			0.44

$$s = \sqrt{\frac{0.44}{14}} = 0.18$$

ν = 7 degrees of freedom

Example B: Estimation from Duplicates, Different Materials

X_f	X_s	$\mid d \mid$	d^2
14.7	15.0	0.3	0.09
20.1	19.8	0.3	0.09
12.5	13.0	0.5	0.25
23.6	23.3	0.3	0.09
15.1	14.9	0.2	0.04
18.2	18.0	0.2	0.04
20.7	20.9	0.2	0.04
Σ			0.64

$$s = \sqrt{\frac{0.64}{14}} = 0.17$$

$$\nu = 7 \text{ degrees of freedom}$$

Estimation of Standard Deviation from the Range

The standard deviation is related to the range, R, of a set of measurements, defined as the difference between the highest and lowest values obtained. The average range, $\bar{R}$, based on several sets (k) of measurements is calculated, and the standard deviation estimate is calculated, using the factor d^*_2 found in a table such as Table C.1 in Appendix C. The calculation is as follows:

$$\bar{R} = \frac{R_1 + R_2 + \ldots + R_n}{n}$$

$$s = \bar{R}/d^*_2$$

The value for d^*_2 will depend on the number of sets, k, used to calculate $\bar{R}$ and the number of measurements in a set. The table also shows the number of degrees of freedom for the estimate of the standard deviation. The following examples will illustrate the calculation.

Example: from the Range of Duplicate Measurements

First Result	Second Result	Range, R
14.5	14.2	0.3
14.8	14.9	0.1
14.05	14.3	0.25
14.2	14.8	0.6
14.9	14.9	0.0
14.3	14.4	0.1
14.7	14.1	0.6
14.4	14.7	0.3
14.1	14.25	0.15
14.95	14.65	0.3
14.25	14.95	0.7

$$\bar{R} = \frac{0.3 + 0.1 + 0.25 + 0.6 + 0 + 0.1 + 0.6 + 0.3 + 0.15 + 0.3 + 0.7}{11}$$

$$\bar{R} = 0.309$$

$$s = \bar{R}/d^*_2$$

d^*_2 = 1.16 for 11 estimates of R (from Table C.1 by interpolation)

$$s = 0.27$$

ν = 10 degrees of freedom for s (from Table C.1, rounded)

Pooling Estimates of Standard Deviations

Several estimates of the standard deviation may be pooled to obtain a better estimate based on an increased number of degrees of freedom. Given several estimates of the standard deviation obtained on several occasions, with the corresponding number of measurements:

$$\begin{array}{cccccc} s_1 & n_1 & \nu_1 & = & \nu_1 - 1 \\ s_2 & n_2 & \nu_2 & = & \nu_2 - 2 \\ \cdot & \cdot & \cdot & & \cdot \\ \cdot & \cdot & \cdot & & \cdot \\ \cdot & \cdot & \cdot & & \cdot \\ s_k & n_k & \nu_k & = & \nu_k - k \end{array}$$

$$s_{pooled} = \sqrt{\frac{\nu_1 s^2{}_1 + \nu_2 s^2{}_2 + \ldots \nu_k s^2{}_k}{\nu_1 + \nu_2 + \ldots + \nu_k}}$$

s_{pooled} will be based on $(\nu_1 + \nu_2 + \ldots + \nu_k)$ degrees of freedom.

Example: Pooling Standard Deviations

The standard deviation of a measurement process was estimated on five occasions. These are to be pooled to improve the estimate of σ.

Trial	s	n	ν
1	0.171	7	6
2	0.205	5	4
3	0.185	7	6
4	0.222	4	3
5	0.180	5	4

$$s_p = \sqrt{\frac{6\times(0.171)^2 + 4\times(0.205)^2 + 6\times(0.185)^2 + 3\times(0.222)^2 + 4\times(0.180)^2}{6 + 4 + 6 + 3 + 4}}$$

s_p = 0.190 estimated with 23 degrees of freedom.

The Analysis of Variance

The total variance which is observed is often the sum of the contributions from a number of sources. Thus,

$$s^2_T = s^2_1 + s^2_2 + \ldots + s^2_n$$

In many measurement situations, the total variance may be considered to arise essentially from two major sources, such as that of the measurement process and that of the samples measured. Within and between days in the same laboratory and within and between laboratories are other examples. By an appropriate experimental design, each variance may be estimated.

As an example, replicate measurements may be made on a number of samples randomly selected from a material of interest. The variation of the replicate measured values on the same sample may be pooled and used to estimate the standard deviation of the measurement process, s_m. The variation of the means of the measured values for the several samples may be used to estimate a total standard deviation, s_T. The standard deviation between samples, s_b (sample homogeneity), may be estimated as follows:

$$s_b = \sqrt{(s^2_T - s^2_m/n)}$$

where n is the number of replicate measurements on each sample.

This simple analysis of variance is discussed further in Chapter 25 in connection with evaluating laboratory performance.

DO TWO ESTIMATES OF PRECISION DIFFER?

The significance of apparent differences in two estimates of a standard deviation is of interest in such situations as:

- possible changes in precision when analyzing control chart data
- whether two methods differ in their precisions
- whether one analyst/instrument/laboratory produces more precise results than another

The statistical significance of apparent differences can be judged on the basis of an F test, conducted as follows:

Let s_1 = estimate of standard deviation (larger value) based on n_1 measurements.
Let s_2 = estimate of standard deviation (smaller value) based on n_2 measurements.

In each case, the respective degrees of freedom, $\nu = n - 1$.

Calculate $F = s^2_1/s^2_2$

Look up the critical value of F_c in Table C.3, based on the respective degrees of freedom for the estimates of s_1 and s_2.

- If $F > F_c$, consider $s_1 > s_2$ at the chosen level of confidence
- If $F < F_c$ there is no reason to believe that $s_1 > s_2$ at the chosen level of confidence

Example: Comparison of Precision Estimates

$s_1 = 2.00$ $\quad n_1 = 6 \quad \nu_1 = 5$
$s_2 = 1.00$ $\quad n_2 = 6 \quad \nu_2 = 5$
$F = 4.00/1.00 = 4.0$
$F_c = 7.15$ at 5% level of significance

Conclusion: There is no reason to believe that $s_1 > s_2$.
Note: If a comparison is to be made with a standard value for a standard deviation, the latter may be considered to be based on an infinite number of degrees of freedom.

WHAT ARE THE CONFIDENCE LIMITS FOR AN ESTIMATE OF A STANDARD DEVIATION?

The width of the confidence interval for an estimated standard deviation will depend on the number of degrees of freedom, ν, upon which the estimate is based ($\nu = n - 1$). The interval is not symmetrical (see Figure 4.3), as in the case for a mean, since a small number of measurements tend to underestimate the standard deviation. In other words, the standard deviation may be somewhat smaller than the estimated value, but it could also be considerably larger. To calculate the bounds of the interval, one may use a table such as Table C.4 and find the factors B_U and B_L corresponding to the number of degrees of freedom involved and the confidence level sought. In the Table, $\alpha = 0.05$ corresponds to a confidence of 95% for the interval so calculated. The confidence interval is then sB_L to sB_U.

Example: Confidence Limits for Estimate of a Standard Deviation

$s = 0.15 \quad n = 10, \quad \nu = 9$

For $\alpha = 0.05$, $\nu = 9$ (see Table C.4) one finds

$$B_U = 1.746; \qquad B_L = 0.6657$$

The confidence interval for s is 0.15 × 0.6657 to 0.15 × 1.746 or 0.10 to 0.26.

CONFIDENCE INTERVAL FOR A MEAN

This is one of the most common statistical calculations. Often, one would like to know with a given probability where the population mean lies with respect to a sample mean. The confidence interval calculated as below essentially does this, in that intervals, so calculated, are expected to include the population mean the selected percentage of times. However, one cannot say whether any specific interval does or does not include the population mean, and this should always be kept in mind. The confidence interval for the mean, $\bar{X}$, will depend on the number of measurements, n, the standard deviation, s, and the level of confidence desired. The confidence interval is calculated using the expression

$$\bar{X} \pm \frac{ts}{\sqrt{n}}$$

The value for t (see Table C.3) will depend on the level of confidence desired and the number of degrees of freedom, ν, associated with the estimation of s. If s is based on the set of measurements used to calculate the mean, $\bar{X}$, then $\nu = n-1$. If the measurements are made by a system under statistical control, as demonstrated by a control chart, ν will depend on the number of measurements made to establish the control limits.

Example: Confidence Interval Based on s
Estimated from a Data Set of Seven Measurements

$$\bar{X} = 10.05$$
$$s = 0.11$$
$$n = 7$$
$$\nu = 6$$

For a 95% level of confidence, t = 2.447, hence:

$$\bar{X} = 10.05 \pm \frac{2.447 \times 0.11}{\sqrt{7}} = \pm\ 0.10 \text{ or } 9.95 \text{ to } 10.15$$

Example: Confidence Interval Based on s
Obtained from Control Chart Limits, One Measurement of x

$$x = 10.05$$
$$s = 0.11$$
$$\nu = 45 \text{ (control chart)}$$
$$n = 1$$

For a 95% level of confidence and 45 degrees of freedom, t = 2.016, hence

$$\bar{X} = 10.05 \pm \frac{2.016 \times 0.11}{\sqrt{1}} = 10.05 \pm 0.22 \text{ or } 9.83 \text{ to } 10.27$$

The values for t were obtained from Table C.3.
Note: There is no statistical basis for a confidence level statement for one measurement unless supported by a control chart or other evidence of statistical control.

DOES A MEASURED VALUE DIFFER FROM AN EXPECTED VALUE?

The analyst may want to know whether a measured value differs significantly from an expected or standard value. The latter could be a specified value for a product, and the significance of the difference is needed when deciding whether to accept or reject it. Or, does the difference of a measured value from a certified value for a reference material indicate that the measurement methodology is biased? In order to answer such questions, the confidence interval for the measured value is calculated. If this includes the certified value, there is no reason to believe that the measured value is in disagreement with the certified value at the level of confidence for the interval.

Example

NBS SRM 2682 Subbituminous Coal was analyzed in triplicate for its sulfur content (certified value S = 0.47%) with the following results:

$$\bar{X} = 0.485 \text{ \% S}$$
$$s = 0.0090 \text{ \% S}$$
$$n = 3$$

Choose the confidence level for the interval, say 95%.

$$\bar{X} = 0.485 \pm \frac{4.303 \times 0.0090}{\sqrt{3}}$$
$$= 0.485 \pm 0.0223$$
$$= 0.463 \text{ to } 0.507$$

Since the certified value lies within the interval, there is no reason to believe that the measured value disagrees with the certified value.

If the measurements had been made while statistical control was demonstrated by a control chart, for which $s = 0.0090$ based on $\nu = 20$ degrees of freedom:

$$\bar{X} = 0.485 \pm \frac{2.086 \times 0.0090}{\sqrt{3}}$$

$$\bar{X} = 0.485 \pm 0.0108 \text{ or } 0.474 \text{ to } 0.496$$

The certified value is not within the interval, hence there is reason to believe that the measured value exceeds the certified value.

DO THE MEANS OF TWO MEASURED VALUES DISAGREE SIGNIFICANTLY?

Questions of this nature arise when the results reported by two laboratories or obtained by two different methods are compared, for example. The decision on disagreement is based on whether the difference, Δ, of the two values exceeds its statistical uncertainty, U. The method used for calculation of the uncertainty depends on whether or not the respective standard deviation estimates may be considered to be significantly different.

Case I. No reason to believe that the standard deviations differ, e.g., same method, analyst, experimental conditions, etc.

Step 1. Choose the significance level of the test.

Step 2. Calculate a pooled standard deviation from the two estimates to obtain a better estimate of the standard deviation.

$$s_p = \sqrt{\left(\frac{\nu_A s^2_A + \nu_2 s^2_B}{\nu_A + \nu_B}\right)}$$

where s_p will be based on $\nu_A + \nu_B$ degrees of freedom

Step 3. Calculate the uncertainty, U, of the difference.

$$U = ts_p \sqrt{\left(\frac{n_A + n_B}{n_A n_B}\right)}$$

Step 4. Compare $\Delta = |\bar{X}_A - \bar{X}_B|$ with U. If $\Delta \leq U$, there is no reason to believe that the means disagree.

Example

$$\bar{X}_A = 4.25 \qquad \bar{X}_B = 4.39$$
$$s_A = 0.13 \qquad s_B = 0.17$$
$$n_A = 7 \qquad n_B = 10$$
$$\nu_A = 6 \qquad \nu_B = 9$$

$$\Delta = | \bar{X}_A - \bar{X}_B | = | 4.25 - 4.39 | = 0.14$$

Step 1. $\alpha = 0.05$ (95% confidence)

Step 2. $$s_p = \left[\frac{6(0.13)^2 + 9(0.17)^2}{6 + 9}\right] = 0.155$$

Step 3. $$U = 2.131 \times 0.155 \ \frac{7 + 10}{70} = \pm\ 0.16$$

Step 4. $0.14 < 0.16$

Conclude that there is no reason to believe that the measured values differ at the 95% level of confidence.

Case II. Reason to believe that the standard deviations differ, e.g., different experimental conditions, different laboratories, etc.

Step 1. Choose α, the significance level of the test.
Step 2. Compute the estimated variance of each value using the individual estimates of the standard deviations.

$$V_A = \frac{s^2{}_B}{n_A} \qquad V_B = \frac{s^2{}_B}{n_B}$$

Step 3. Compute the effective number of degrees of freedom, f*.

$$f^* = \frac{(V_A + V_B)^2}{\dfrac{V_A}{n_A + 1} + \dfrac{V_B}{n_B + 1}} - 2$$

Step 4. Compute the uncertainty, U, of the difference. $U = t' \ (V_A + V_B)$, where t' is the effective value of t based on f* degrees of freedom
Step 5. Compare Δ with U. If Δ is $\leq$ U there is no reason to believe that the means disagree.

Example

$$\bar{X}_A = 4.25 \qquad \bar{X}_B = 4.39$$
$$s_A = 0.13 \qquad s_B = 0.17$$
$$n_A = 7 \qquad n_B = 10$$

Step 1. $\alpha = 0.05$ (95% confidence)

Step 2. $V_A = \frac{(0.13)^2}{7} = 2.414 \times 10^{-3}$ $\quad V_B = \frac{(0.17)^2}{10} = 2.89 \times 10^{-3}$

Step 3. $$f^* = \frac{(2.414 \times 10^{-3} + 2.89 \times 10^{-3})^2}{\frac{2.414 \times 10^{-3}}{8} + \frac{2.89 \times 10^{-3}}{11}} - 2$$

$f^* = 17$

Step 4. $U = 2.11 \quad (2.414 \times 10^{-3} + 2.89 \times 10^{-3})$

$U = 0.153$

Step 5. $\Delta = 0.14; \quad U = 0.153$

Conclude there is no reason to believe that $\bar{X}_B > \bar{X}_A$ at 95% level of confidence.

STATISTICAL TOLERANCE INTERVALS

A tolerance interval represents the limits within which a specified percentage of the population is expected to lie with a given probability. It is especially useful to specify the limits of variability in composition of samples—that is to say, the limits within which a specified percentage of samples of material are expected to lie. If the standard deviation of the population of samples were known, the limits for a given percentage of the population could be calculated with certainty. Because only an estimate of the standard deviation is usually known, based on a limited sampling of the population, a tolerance interval based on inclusion of a percentage of the population with a specific probability of inclusion is all that can be calculated.

The calculation is made as follows:

$$\text{Tolerance Interval} = \bar{X} \pm ks$$

where k = a factor (obtained from Table C.5 for example) based on the percentage, p, of population to be included, the probability, τ, of inclusion, and the number of measurements used to calculate $\bar{X}$ and s.

Example: Statistical Tolerance Interval

Based on ten measurements, the sulfur content of a shipment of coal was found to be:

$$\bar{X} = 1.62\%\ S \qquad s = 0.10\%\ S \qquad n = 10$$

From Table C.5, $k = 3.379$ for $\tau = 0.95$, $p = 0.95$, $n = 10$. The tolerance interval is thus 1.62%, $\pm 0.34\%$ or 1.28% to 1.96% S.

POOLING MEANS TO OBTAIN A GRAND AVERAGE, $\bar{\bar{X}}$

There are various occasions when one might desire to combine data sets. An example is the computation of a consensus value, based on the values reported by several laboratories. The statistical principles for use in averaging compatible data (see Chapter 22) are given below (see also Ku [82]).

Case I. All means based on same number of measurements of equal precision. Calculate a simple grand average as follows:

$$\bar{\bar{X}} = \frac{\bar{X}_1 + \bar{X}_2 + \ldots + \bar{X}_n}{n}$$

Case II. Means based on different numbers of measurements, but no reason to believe the precisions differ. Calculate a number weighted average as follows:

$$\bar{\bar{X}} = \frac{n_1\bar{X}_1 + n_2\bar{X}_2 + \ldots + n_n\bar{X}_n}{n_1 + n_2 + \ldots + n_n}$$

Case III. Means based on different numbers of measurements and/or with differing precisions. Calculate an average, weighted inversely according to the respective variances of the means, as follows:

Step 1. Compute weight to be used for each mean.

$$w_i = \frac{n_i}{s^2_i} \qquad w_1 = \frac{n_1}{s^2_1}$$

Step 2. $$\bar{\bar{X}} = \frac{w_1\bar{X}_1 + w_2\bar{X}_2 + \ldots + w_n\bar{X}_n}{w_1 + w_2 + \ldots + w_n}$$

$$s^2_{\bar{\bar{X}}} = \frac{1}{w_1 + w_2 + \ldots + w_n}$$

Example: Calculation of Grand Average, Case I

Calculate the grand average, $\bar{\bar{X}}$, of the following means, all believed to be equally precise.

$$\bar{X}_1 = 10.50$$
$$\bar{X}_2 = 10.37$$
$$\bar{X}_3 = 10..49$$
$$\bar{X}_4 = 10.45$$
$$\bar{X}_5 = 10.47$$

$$\bar{\bar{X}} = \frac{10.50 + 10.37 + 10.49 + 10.45 + 10.47}{5}$$

$$\bar{\bar{X}} = 10.456$$

Example: Calculation of Grand Average, Case II

$$\bar{X}_1 = 10.50 \qquad n = 10$$
$$\bar{X}_2 = 10.37 \qquad n = 5$$
$$\bar{X}_3 = 10.49 \qquad n = 20$$
$$\bar{X}_4 = 10.45 \qquad n = 5$$
$$\bar{X}_5 = 10.47 \qquad n = 7$$

$$\bar{\bar{X}} = \frac{10.50 \times 10 + 10.37 \times 5 + 10.49 \times 20 + 10.45 \times 5 + 10.47 \times 7}{47}$$

$$\bar{\bar{X}} = 10.472$$

Example: Calculation of Grand Average, Case III

i	$\bar{X}_i$	n_i	s_i	w_i
1	10.50	10	.10	1000
2	10.37	5	.15	222
3	10.49	20	.11	1652
4	10.45	5	.10	500
5	10.47	7	.16	273

$$\bar{\bar{X}} = \frac{10.50 \times 1000 + 10.37 \times 222 + 10.49 \times 1652 + 10.45 \times 500 + 10.47 \times 273}{1000 + 222 + 1652 + 500 + 273}$$

$$\bar{\bar{X}} = 10.478$$

$$s_{\bar{\bar{X}}} = 0.0166$$

OUTLIERS

Outliers are essentially individuals that do not belong in a population, or there is less than a specified probability that they do. The concept applies equally to samples of material as well as to individual measurements, and even to whether the data outputs of one laboratory in a group are different from the others in the group. The following discussion is concerned with questions about data, but it should be obvious how the ideas would be extended to the other situations.

Outliers in data sets can result from such causes as blunders or malfunctions of the methodology, or from unusual losses or contamination. If outliers occur too often, this could indicate that there may be deficiencies in the quality control program which can be corrected and thus improve the measurement process. One should always search diligently for causes before data are rejected.

Possible outliers can be identified when data are plotted, when results are ranked, and when control limits are exceeded. Only when a measurement system is well understood and the variance is well established, or when a large body of data is available, is it possible to distinguish between extreme values and true outliers with a high degree of confidence.

The following rules for rejection of data should be used with caution since an outlier in a well-behaved measurement system should be a rare occurrence.

Rejection for Assignable Cause

Whenever an outlier is suspected, the analyst should look for a reason for its occurrence—that is to say, an assignable cause. Miscalculations, use of wrong units, system malfunctions, misidentification of a sample, transcription errors, and contamination are examples. In some cases, it may be possible to take corrective actions and salvage what would otherwise be a bad measurement. Records should be kept of such problems and the corrective actions taken, for guidance in possibly modifying quality control procedures.

Rule of the Huge Error

If the questioned value, X_q, differs from the mean by an appropriate multiple, M, of the known or assumed standard deviation, s, it may be considered to be an outlier. The size of the multiple depends on the confidence required for rejection. One evaluates:

$$M = \frac{|X_q - \bar{X}|}{s}$$

A practical rule might be to use $M \geq 4$ as a criterion for rejection. This corresponds to a significance level of $< 2\%$ when the standard deviation is well established, such as based on a data set of 15 or larger.

If s is not well established but depends on the data set in question, the odds for rejection are much larger. For example, if $\bar{X}$ and s are based on 6 measurements, $M > 6$ would be the criterion for rejection for a 2% level of significance.

Example: Outliers, Huge Error Concept

X (Original Order)	X (Ranked Order)
10.50	10.45
10.47	10.47
10.49	10.47
10.45	10.48
10.47	10.49
10.57	10.50
10.52	10.50
10.50	10.52
10.48	10.53
10.53	10.57

1. 10.57 appears to be an outlier.
2. Calculate mean and sample standard deviation, s, ignoring 10.57.

$$\bar{X} = 10.490 \qquad s = 0.0255$$

$$M = \frac{|10.57 - 10.49|}{0.0255} = 3.13$$

3. Since 3.13 < 4 conclude that 10.57 should be retained in the data set.
4. Recalculate mean and sample standard deviation including the value 10.57.

$$\bar{X} = 10.498 \qquad s = 0.0349$$

Statistical Tests

Several statistical tests are available for confirming a suspicion of outliers, based on ranking the data and testing extreme values for credibility. The Dixon test and the Grubbs test are used widely and described in the following sections.

Dixon Test for Outlying Observations

This procedure has the advantage that an estimate of the standard deviation is not needed to use it. The procedure is as follows:

1. Rank the data in order of increasing numerical value, i.e.,

$$X_1 < X_2 < X_3 < \ldots < X_{n-1} < X_n$$

2. Decide whether the smallest X_1, or the largest, X_n, is suspected to be an outlier.
3. Select the risk you are willing to take for false rejection.
4. Compute one of the following ratios (statistic):

n	Ratio	If X_n is Suspect	If X_1 is Suspect
$3 \leq n \leq 7$	τ_{10}	$(X_n - X_{n-1})/(X_n - X_1)$	$(X_2 - X_1)/(X_n - X_1)$
$8 \leq n \leq 10$	τ_{11}	$(X_n - X_{n-1})/(X_n - X_2)$	$(X_2 - X_1)/(X_{n-1} - X_1)$
$11 \leq n \leq 13$	τ_{21}	$(X_n - X_{n-2})/(X_n - X_2)$	$(X_3 - X_1)/(X_{n-1} - X_1)$
$14 \leq n \leq 25$	τ_{22}	$(X_n - X_{n-2})/(X_n - X_3)$	$(X_3 - X_1)/(X_{n-2} - X_1)$

5. Compare the ratio (statistic) calculated with the values in Table C.7. If the calculated ratio is greater than the tabulated value, rejection may be made with the tabulated risk.

Example

Given the ranked data set:

10.45, 10.47, 10.47, 10.48, 10.49, 10.50, 10.50, 10.52, 10.53, 10.58

The value 10.58 is suspect.

1. Calculate τ_{11}.

$$\tau_{11} = \frac{10.58 - 10.53}{10.58 - 10.47} = \frac{0.05}{0.11} = 0.454$$

2. At 5% risk of false rejection (Table C.7), $\tau_{11} = 0.477$.
3. Conclusion: no reason to reject the value 10.58.

Grubbs Test for Outlying Observations

This test is useful for making statistical decisions on the rejection of outliers. The procedure for using it is as follows:

1. Rank the data in order of increasing numerical value.

$$X_1 < X_2 < X_3 < \ldots < X_{n-1} < X_n$$

2. Decide whether the smallest, X_1, or the largest, X_n, is suspected to be an outlier.
3. Estimate the standard deviation, s, of the data set (using all data).
4. Calculate the appropriate value of T as follows:

$$T = \frac{\bar{X} - X_1}{s} \quad \text{or} \quad T = \frac{X_n - \bar{X}}{s}$$

5. Select the risk you are willing to take for false rejection.
6. Compare T with values tabulated in Table C.8, depending on n and the acceptable risk. If T is larger than the tabulated value, rejection may be made with the associated risk.

Example

Given the ranked data set:

10.45, 10.47, 10.47, 10.48, 10.49, 10.50, 10.50, 10.52, 10.53, 10.57

The value 10.57 (the largest value) is suspect.

1. Calculate $\bar{X}$ and s.

$$\bar{X} = 10.499 \qquad s = 0.034$$

2. Calculate T.

$$T = \frac{10.57 - 10.499}{0.0341} = 2.082$$

3. Compare T with values in Table C.8.

At 5% risk, for $n = 10$, $T = 2.176$

4. Conclusion: no reason to reject 10.57.

Youden Test for Outlying Laboratories

Youden [155] describes a statistical test applicable to a group of laboratories to identify ones that consistently report high or low results. All laboratories are given identical samples on a number of occasions and the results reported are ranked in descending order, i.e., $X_1 > X_2 \ldots > X_n$. The laboratory reporting X_1 receives a score of 1 and so forth. The scoring is done on each occasion, and the cumulative score of each laboratory is computed. The range for the cumulative scores, due to chance and based on 95% confidence, are listed in Table C.9. Laboratories whose cumulative scores are outside of these limits are considered to be consistent outliers. If the cumulative score is smaller than the range in the table, the reported results are considered to be consistently high; if larger, the reported results are considered to be consistently low.

Example

	Sample Results				
Laboratory	1	2	3	4	5
A	10.5	14.2	20.0	18.1	12.3
B	9.9	13.7	19.7	18.2	11.7
C	10.2	14.1	19.9	17.8	12.0
D	9.7	13.9	19.5	17.9	12.2
E	10.4	14.0	19.7	17.5	11.6
F	10.0	13.6	19.4	17.6	11.9
G	10.1	13.8	19.6	17.7	12.1

	Sample Scores					Cumulative
Laboratory	1	2	3	4	5	Scores
A	1	1	1	2	1	6 low
B	6	6	3	1	6	22
C	3	2	2	4	4	15
D	7	4	6	3	2	22
E	2	3	4	7	7	23
F	5	7	7	6	5	30
G	4	5	5	5	3	22

For seven laboratories reporting on 5 samples, the range is expected to lie within 8 to 32, with 95% confidence. Accordingly, laboratory A is considered to provide results higher than other members of the group.

Cochran Test for Extreme Values of Variance

A test described by Cochran may be used when deciding whether a single estimate of a variance is extreme (excessively large) in comparison with a group with which it is expected to be comparable. For example, is the variance reported by one laboratory excessively large compared to those of other members of a group of laboratories? All of the variances of interest are summed, i.e., Σs^2_i. If the ratio, s^2 (largest)/Σs^2_i, exceeds the value for Cochran's test listed in Table C.10, the largest variance is considered to be an extreme value (an outlier).

STATISTICS OF CONTROL CHARTS

The preceding discussion was illustrated for situations related to the interpretation of sets of measurement data. The use of statistics in establishing and/or revising control limits is similar to the above, but several aspects will be discussed in the following sections in the interest of completeness.

Control Limits

Control limits are based upon estimates of standard deviations which may be obtained from a series of measurements or from the range of sets of measurements. From time to time, it may be necessary to decide whether the precision of the measurement system has changed, thus requiring revision of the control limits. This decision involves the application of the F test described on page 26. Because this test is unable to distinguish between small differences in the case of small data sets, it is recommended that each of the s values in question be based on at least 14 degrees of freedom. If the precision has changed significantly, new limits will need to be calculated, based on the latest data. If not, control limits should be revised by combining data bases by pooling, as described on page 24.

Central Line Considerations

Agreement with Established Value

The question may arise as to whether the value for $\bar{X}$, the mean of a series of measurements on a control sample, differs significantly from an expected or certified value for the sample. The confidence interval for the mean is calculated first, as explained earlier. If this interval includes the expected value, there is no reason to believe, at the confidence level selected, that $\bar{X}$ differs from it. In this case, the certified value is used as that for the central line. If there is a significant difference, then $\bar{X}$ is considered to be biased, and an assignable cause should be sought. The experimental value for the mean may be used as the central line until such time as the cause for the bias is resolved and the certified value can be used.

Has the Central Line Changed?

The central line may be based on a measured mean, $\bar{X}$. At a later time, after accumulation of additional data, the question may arise as to whether the mean of the data accumulated since a previous determination differs significantly from the former value. The decision is based on the significance of differences of means discussed on page 30. The method of calculation will depend on whether the two estimates of the respective standard deviations may be considered to be significantly different as discussed earlier.

Should the means be found to disagree significantly, the central line should be based on the latest data. Otherwise the means are combined as described on page 31, Case II.

USE OF RANDOM NUMBER TABLES

It is often desirable to randomize the sequence in which measurements are made, samples are chosen, and other variables of an analytical program are set. Tables of random numbers, such as Table C.10 are a convenient and simple way to accomplish this. The following procedure may be used.

1. Number the samples (or measurements) serially, say 00 to xy. For example, 00 to 15 for 16 items.
2. Start at any randomly selected place in the table and proceed from that point in any systematic path. The order in which the item numbers are located becomes the random sequence number to be assigned to them.

Example

Start at Row 7, Column 3 (chosen by chance) of Table C.10 and proceed from left to right as in reading. The first number is 39 which is not usable for the above series of items. The first usable number is 10. Proceeding as above, the items are located in the following order: 10, 08, 01, 05, 06, 04, 00, 07, 03, 11, 09, 15, 02, 13, 12, 14. If a number already chosen is encountered, pass over it to the next usable number.

CHAPTER 5

Chemical Analysis as a System

A fundamental premise of quality assurance is that measurement may be established as a process, analogous to a manufacturing process. The process may be brought into a state of statistical control, i.e., the individual measurements are defined by a statistical distribution, and its characteristic precision and accuracy can be assigned to the data output. Obviously, quality control relates to all that is done to attain and maintain the state of statistical control.

SYSTEMATIC APPROACH TO ANALYSIS

With only a little reflection, it is clear that chemical analysis is more than a single process and may be better characterized as a system consisting of a series of interdependent processes as indicated by the flow diagram of Figure 5.1. Chemical measurements are undertaken to answer questions necessary in solving problems. A problem must be represented by a model that sets forth the questions that need to be answered and defines the data requirements and their utilization. Based on the model, a measurement program may be planned which addresses sampling, calibration, the methodology to be used, and the quality assurance procedures to be followed. The design and implementation of the chemical measurement system is often an iterative process, and every measurement program should include a feedback mechanism which provides for identification of deficiencies and the initiation of corrective actions for all parts of the system as required.

All components of the system are critically dependent upon each other and are implicitly or explicitly present in every measurement activity. Failure to recognize the importance of each component and to properly design an appropriate system is a major defect of many analytical measurement programs.

EFFECTIVE PROBLEM SOLVING

For effective solution of a problem, the conditions shown in Figure 5.2 must be satisfied. The model must be correct, or else the data can hardly be useful

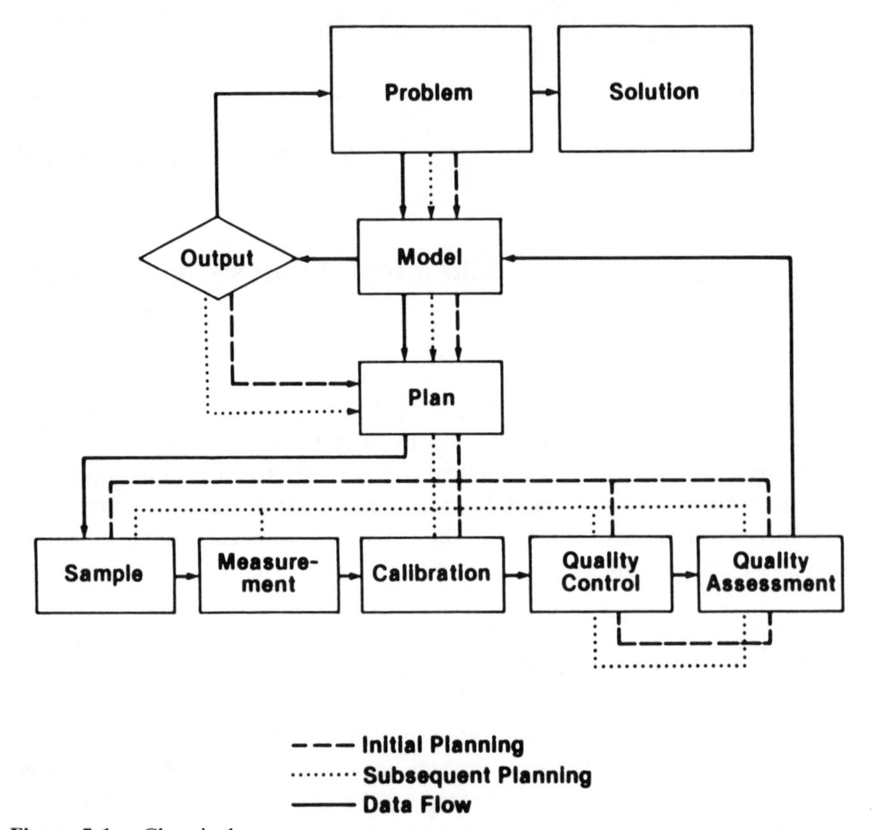

Figure 5.1. Chemical measurement system.

for the intended purpose. Since the plan addresses all technical aspects of the measurements, it is obvious that it cannot have any major faults. The need for relevant samples and appropriate methodology is hardly debatable. Inadequacies in the calibration process could result in measurement bias that would vitiate the data. Quality assurance must be practiced properly to ensure evaluated data.

All of the nouns in Figure 5.2 are preceded by modifying adjectives that have been chosen with considerable thought. However, there are no absolute criteria to judge whether the conditions that they imply have been met beforehand. Rather, professional judgment and past experience are the only avenues available to guide one in the design of an effective analytical system to solve a given problem. The succeeding chapters have been included to review the principles of good measurement and to provide guidance in the design of reliable measurement systems.

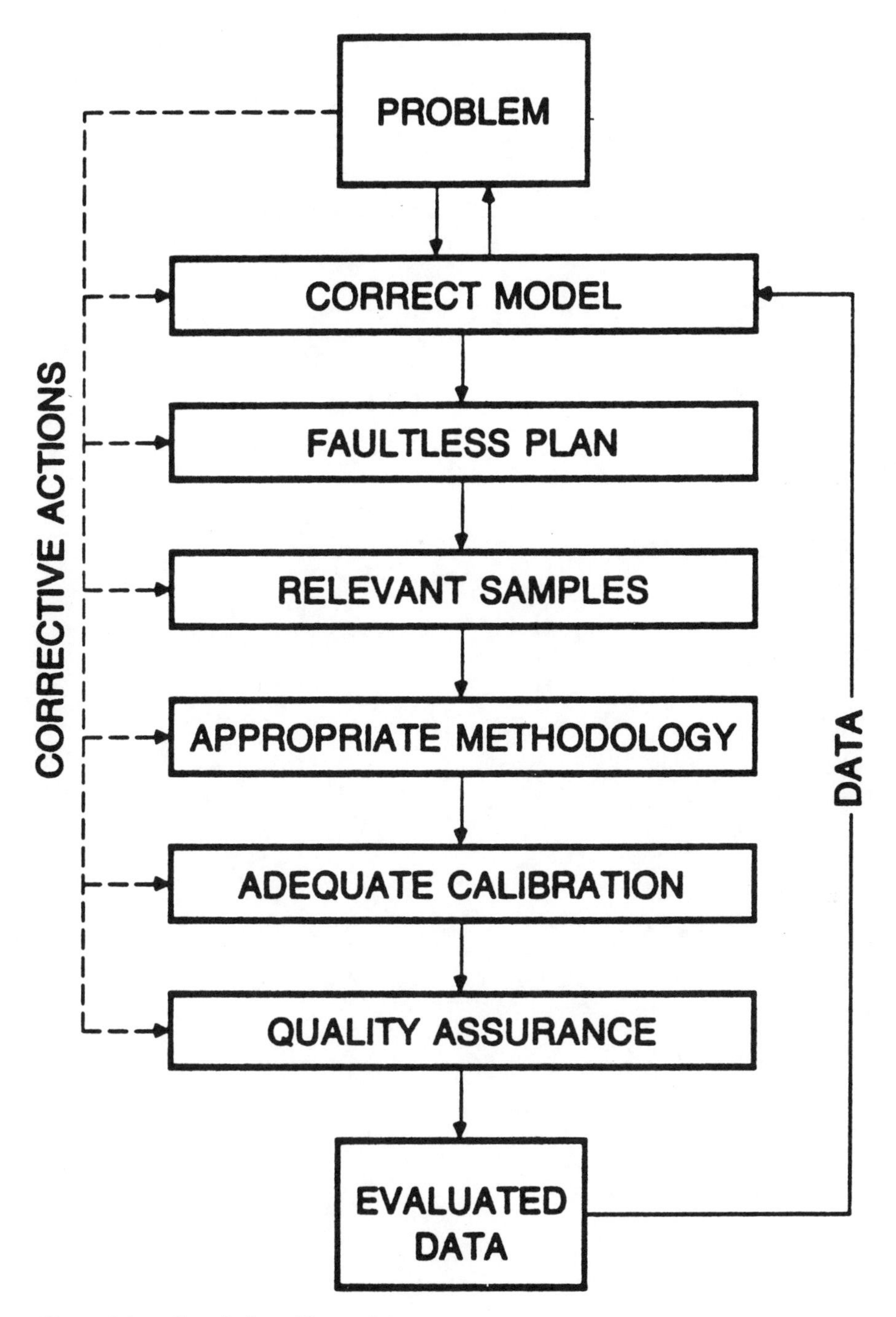

Figure 5.2. Chemical problem solving.

CHAPTER 6

The Model

A model is explicitly or implicitly involved in every attempted analytical solution of a problem. The model may be theoretical, such as the ideal gas law, or empirical, as in the determination of total dissolved matter. In the latter case, it may be decided arbitrarily that any material that passes through a 0.4 μm filter will be considered to be dissolved and any unfilterable residue as undissolved. The model thus becomes the 0.4 μm cutoff point, and whether or not the model is followed in any measurement program will depend on the availability and use of the designated filter and how well accurate phase separations can be achieved by the filtration process. Clearly, the use of any other filter would constitute a different model of the problem, and the data obtained in its use would not necessarily be equivalent or even relatable to that obtained using the specified filter.

Empirical models are based on assumptions, and some measurements are devoted to model testing. If the data fit the model, the latter may be found to be useful for problem solving and even prediction, but its fundamental validity is not necessarily proved.

IMPORTANCE OF A MODEL

The critical relation of the model to the problem should be obvious, and decisionmakers must be aware of the consequences of using an incorrect one. Small errors in the model could result in the measurement of irrelevant samples by inappropriate methodology. Whenever decisions are made, full details of the model must be included so that its correctness can be assessed at any time this is questioned.

Failure to follow the model can result in analytical uncertainties and even useless data of otherwise acceptable quality. When a model is not specified clearly, the analyst may follow his intuition as to what is appropriate. If his concept of the problem should change from time to time, variability of data could result due to such causes alone.

Samples and sampling are especially sensitive to defects of models and to their imprecise specification. This subject is addressed further in Chapter 8.

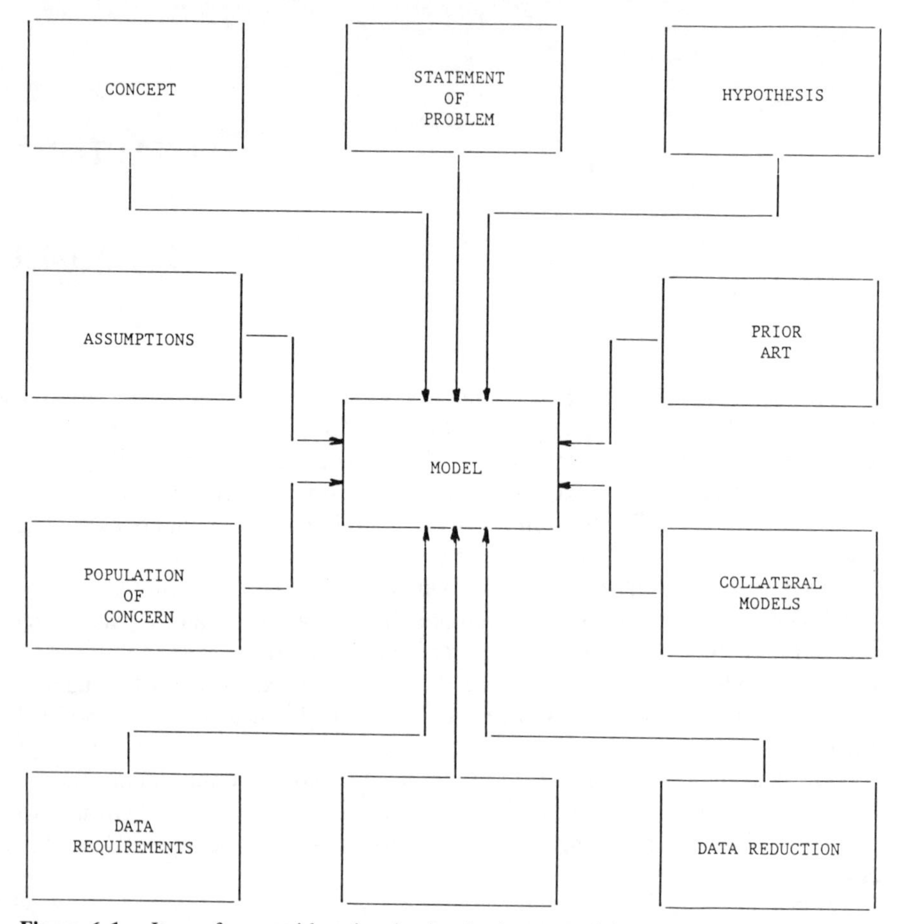

Figure 6.1. Items for consideration in developing a model.

DEVELOPMENT OF A MODEL

The development of an appropriate model requires a thorough understanding of the phenomena or the system under investigation. If there are gaps in this knowledge, there could be defects in the model. In the case of a complex problem, it may need to be subdivided into subproblems that can be individually addressed and modeled. The relation of the whole to the parts needs to be known in such cases.

Figure 6.1 indicates the major items that ordinarily need to be taken into consideration when developing a model for a specific measurement problem. Scientific investigation consists figuratively and even literally in answering well-phrased questions. These questions define the problem, and hence, they must be well defined. Questions related to chemical measurements required for problem solving include:

- statement of problem in as specific terms as possible
- information already known
- simplifying assumptions
- analyte(s) of concern
- concentration level(s) of concern
- speciation of concern: element, compound, complexes
- location of analyte(s) of concern: surface, specific areas, overall
- what is to be determined: total amount/concentration present or some fractional (e.g., available) amount
- phase relations: solid, liquid, gaseous, dissolved, suspended
- interrelationships of analyte(s) of concern
- particle-size distribution
- collateral information that could influence data
- accuracy requirements for data
- mathematical relationships that will be used to relate data to problem

When the model is developed, it should be peer reviewed. This will confirm its soundness or suggest ways in which it might be improved to enhance the ability of the data to solve the problem. Both initial and periodic reviews during its use are recommended to judge its applicability. Thus, model development should be an iterative process, just as is the case for every other aspect of the analytical system.

In the case of adversarial situations, all parties concerned or their technical advisors should agree that the "right" data is being obtained for which a "correct" model is necessary. Unless this is achieved, there can be little hope that anyone will be satisfied with the outcome of a measurement program. In legal proceedings, the opposing attorneys may have pretrial agreement on the issues in a process known as stipulation. Such a process could be used with advantage in many measurement programs.

Because of the direct and often critical relation of the analytical data to the model, the latter must be clearly identified whenever measurement results are reported. Any limitations that the model places on the data need special emphasis.

DATA QUALITY OBJECTIVES

An appropriate model will provide guidance as to the quality of data that will be necessary to solve the problem or to provide useful information for this purpose. Indeed, the setting of realistic quality requirements is a prerequisite for planning a measurement program. Data of insufficient quality will have little value for problem solving while data of excess quality may provide little if any additional information, and may be counterproductive if extra costs are involved. Accordingly data quality objectives must be based on cost-effective considerations based on realistic needs of the problem and the capability of the measurement process. In addition, knowledge of the expected variability of the population sampled is needed to estimate the number of samples to be taken and the number of measurements to be made on each sample. Guidance for making such decisions is presented in later chapters of this book.

Concerning accuracy, qualitative decisions are commonly considered to require a relative accuracy of ± 30 percent; quantitative data used for hypothesis testing should have a minimum accuracy of about ± 10 percent; semiquantitative data will lie somewhere between the above limits.

A factor of three is useful when considering the accuracy requirements for data. Thus, when the uncertainty of a measured value is one-third or less of the permissible tolerance for its use, it can be considered as essentially errorless for that use.

The key consideration in problem solving is a realistic appraisal of quality requirements for the specific situation. Then, analytical methodolgy should be sought that is adequate to provide such quality. As the data quality requirements and the analytical capability approach each other, the resulting numerical data can become utterly meaningless. Unfortunately, much environmental data is taken without due consideration of the above. Indeed, the analytical state of the art has far outclassed that of data interpretation in a number of fields. In many cases, it would seem advisable to measure more samples with less sensitivity than to push down detection levels, requiring expensive measurements and accordingly on fewer samples.

There is no point in looking for nondetectable amounts of any substance in any thing or determining whether any appreciable amount of a substance may be present. Detection limits are fast approaching single atom or single molecule levels. It has been estimated that concentrations of 1 in 10^{18} of almost any substance can be expected to be present in almost any sample. This translates to more than a thousand molecules.

CHAPTER 7

Planning

After, and only after a model has been established, plans can be developed to provide guidance and direction for the various operations of the analytical system. Planning is one of the most important, yet often one of the most poorly executed, steps in chemical measurements. Sometimes this is due to expediency. The need for analytical data may not be anticipated soon enough and, when recognized, there is insufficient or no time to plan. Procrastination on someone's part, ranging from the data user to the data producer, may produce a measurement on the basis of "the best that can be done immediately." Of course, there are real emergency situations, such as an accident that requires some kind of information to direct reasonably prudent actions, yet little if any time is available for any but rudimentary planning. Unfortunately, there are times when planning could have taken place but did not happen because of oversight or ignorance. This should not happen in a well-operated laboratory. Lack of planning can cause problems ranging from disastrous results to data of uncertain quality. Accordingly, the need for adequate planning should be recognized, and it should be a requirement of the quality assurance program.

GENERAL ASPECTS OF PLANNING

The several items that need to be considered when planning a measurement program are indicated in Figure 7.1 and will be elaborated on in the following paragraphs. Planning involves several aspects of the analytical system. Selection of relevant samples and appropriate methodology is obviously involved. The first of these includes the kind, number, location, and mechanics of implementing the sampling plan. The latter involves selection and/or validation of methodology, decisions on calibration, and the measurement program to be followed. Designating the appropriate quality assurance practices and the kind and frequency of measurement of quality assessment samples will be involved. Because a number of options will emerge, a cost-benefit analysis is ordinarily required.

Even a simple measurement will need more or less planning, especially if it is

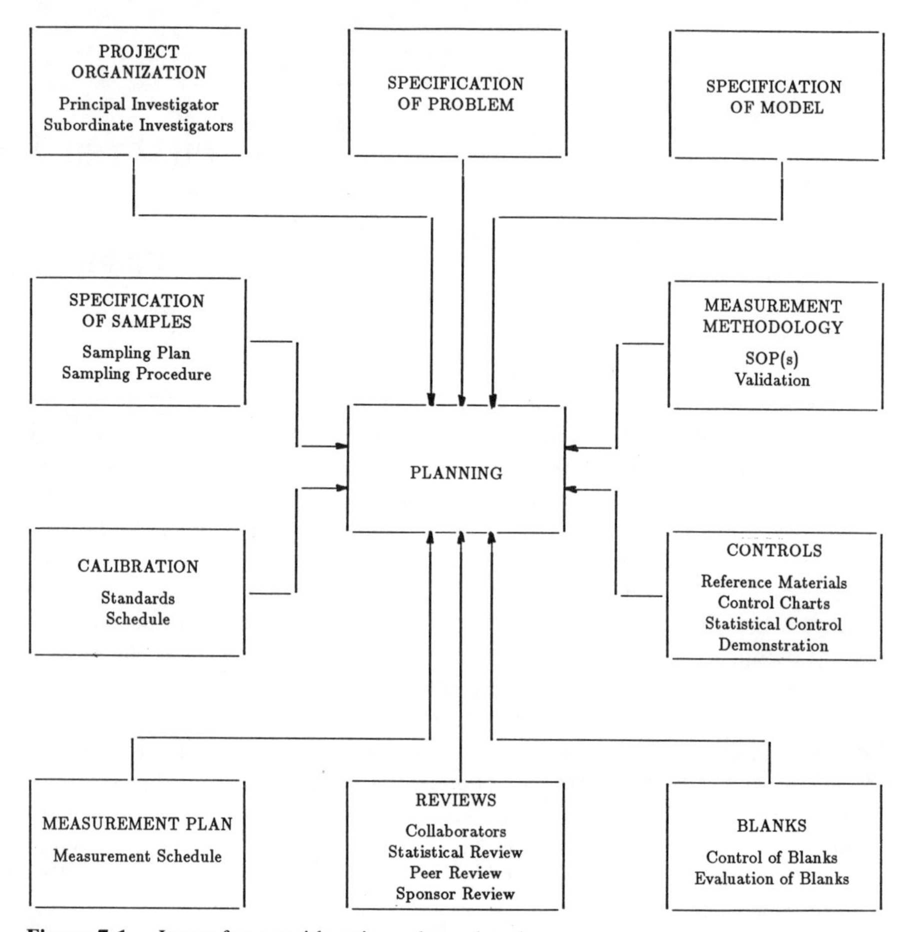

Figure 7.1. Items for consideration when planning a measurement program.

a new or seldom-occurring activity of a laboratory. As the criticality and/or sensitivity of the data increases and when a large and/or complex project is concerned, the amount of planning effort will increase as well. In such cases, a written plan will be required which will address most if not all of the items on the checklist included at the end of this chapter.

Planning should be approached from three points of view: discipline expertise in the field(s) to which the data is related, measurement expertise, and statistical expertise. For small projects, all of the expertise may be in one or several individuals. In complex situations and especially in some regulatory investigations (see Chapter 26), a principal investigator and subinvestigators may be involved. The need for coordination is obvious in such a case.

Peer review of any plan that is developed should be a matter of policy, both to spot faults and to achieve consensus of acceptance by all interested parties. Moreover, the plan may need review periodically during the program of work to ensure its adequacy. Experimental design is always important, especially when the purpose of experimentation is to seek maximum or optimum levels of the experimental factors. Excellent books are available to give guidance in this area. In all but simple programs, it is advisable to consult with a competent statistician who is experienced in experimental design. The design is, of course, only one aspect, albeit a very important part of an experimental plan. Guidelines for development of all aspects of an adequate plan follow.

DEVELOPMENT OF AN EXPERIMENTAL PLAN

The plan is developed by the principal investigator in collaboration with the various participants. In the case of a complex and/or interdisciplinary measurement program, the persons responsible for the individual operations may each develop their parts of the plan which are coordinated by the principal investigator. The statistical aspects of the plan may be developed in collaboration with or may need to be reviewed by a statistician. The plan may need to be reviewed by a sponsor in some cases. Peer review before initiation of work is highly recommended.

In some cases, it may not be possible to develop a definite plan in advance of the work. In such situations, as much as possible should be done initially, followed by preliminary work to finalize it. In any case, the plan should be scrutinized periodically as necessary to verify that it is credible and useful for the measurement undertaken.

The content of the plan is summarized in the following sections. The detail with which each element is addressed will depend upon the complexity of the measurement problem and the critical nature of the data produced.

CONTENT OF A MEASUREMENT PLAN

Project Organization

The plan will specify the principal investigator who will have prime responsibility for implementing it, revising it as necessary, and for initial release of data. The principal investigator will interact with any sponsor through appropriate management channels. The names of the subordinate investigators will be specified together with the parts of the program for which they have responsibility.

Specification of Problem

The nature of the problem to be solved and/or the general work to be done will be described in quantitative terms, as possible. For example, the analytes to be measured, the accuracy requirements, and other details of the aims and objectives should be so specified.

Specification of Sample

The information base that will define the sample(s) should be stated. This will include any sample treatment such as drying, and the size of sample to be used. Some type of subsampling will be involved in most cases, and the procedure for doing this should be described specifically. The procedure to be used for any analytical processing of the sample should be described. The sampling plan should describe how the samples are to be selected. Any procedures needed to protect or preserve the samples during storage and/or measurement should be described. The chain of custody to be followed, if pertinent, should be specified.

Measurement Methodology

The method of measurement and the procedure to be followed should be specified, the latter in the form of a written procedure (SOP). If a procedure in common use is employed, it may be cited, provided it is described elsewhere. If modifications are made of a procedure of record, only these modifications need to be recorded. In all of the above, sufficient detail should be given to be informative to a third party with competence in the measurement area.

The method to be used in validating the methodology, prior to its use, should be stated. The way by which statistical control is to be demonstrated, prior to conducting definitive measurements, should be defined.

Calibration

The details of calibration should be described. This will include the materials to be used, how standards are to be prepared, and how the confidence in them is to be established and quantified. The calibration schedule to be followed should be determined and stated.

Controls

The controls to be used must be described together with the frequency of their use. How control data are to be used in control charts or otherwise should be stated. Whenever possible, an SRM should be at least one of the controls that is used.

Measurement Plan

The measurement plan will describe the checkout plan to verify statistical control, prior to measurement, and the measurement schedule to be followed. The sequence of samples, controls, blanks, and calibrations should be planned and stated.

Blanks

Where the correction for a blank is significant, the way the blank is controlled and the way this is to be verified should be stated. Control charts may be the best way to do this.

ADDITIONAL REMARKS

The development of the measurement plan may be based on statistical design or on professional judgment. In complex programs, the former is often necessary. In other cases, the latter may be adequate. Parts of even simple programs will require some statistical planning since probabilistic statements will need to be made for the limitations of the data. Accordingly, the use of a statistical design is recommended in planning all aspects of the experimental work to the extent possible.

Planning is considered to be essential for reliable work and cost-effective utilization of resources. Written plans that are followed closely provide a record of what was done and add to the defensibility of laboratory outputs. Also, they provide the basis for peer review which can provide valuable input to a measurement process.

The checklist that follows is given to assist in the development of a comprehensive measurement plan.

Planning Checklist

	Yes	No	N.A.
Project Organization			
Principal Investigator	___	___	___
Subordinate Investigator	___	___	___
Specification of Problem	___	___	___
Specification of Model	___	___	___
Specification of Sample(s)			
Sampling Plan	___	___	___
Sampling Procedure	___	___	___
Measurement Methodology			
SOP(s)	___	___	___
Validation	___	___	___
Statistical Control Demonstration	___	___	___
Calibration			
Calibration Standards	___	___	___
Calibration Schedule	___	___	___
Controls			
Reference Materials	___	___	___
Control Charts to be Used	___	___	___
Measurement Plan			
Measurement Schedule	___	___	___
Blanks			
Control of Blanks	___	___	___
Evaluation of Blanks	___	___	___
Reviews			
Review by Collaborators	___	___	___
Statistical Review of Plans	___	___	___
Peer Review of Plans	___	___	___
Supervisory Review	___	___	___
Review by Sponsor	___	___	___

CHAPTER 8

Principles of Sampling

The purpose of most measurements, and especially chemical measurements, is to evaluate some property of a material of interest. The value of the property is then used in a decision process to make judgments on such questions as its suitability for a specific use, the need for some corrective action, or the conformance with some specification. The samples measured are often the most critical aspect of a measurement program. They must be relevant and truly represent the universe concern.

In limited cases, the specimen tested is the universe of concern. For example, in forensic analysis the identification of a single chip of metal could be decisive for settling a legal question. In most cases, the specimens tested are only a part of the universe of concern. Then, it is extremely important to know how the specimens are related to the universe since the decision process is concerned with the universe and not with the specimens tested. Even in the first case where the specimen is the universe of concern, its identity must be unquestionable. Thus, questions about the sample are always crucial for every measurement situation.

The various matters that need to be considered when developing a plan to obtain, prepare, and use adequate samples are shown diagrammatically in Figure 8.1 and discussed in the following sections.

INITIAL CONSIDERATIONS

The first problem is to identify the universe of concern. There must be a clear conception of what is to be sampled; hence, the physical, chemical, and dimensional parameters that define it must be known. One can think of three kinds of universes: (1) single item, in which the entire universe is evaluated; (2) discrete lot, consisting of a finite number of discrete individuals, e.g., items from a production lot; and (3) bulk, a massive material composed of arbitrary and/or irregular units. A "defined bulk" is one in which the boundaries are clearly distinguishable while a "diffuse bulk" is one in which the limits are ill-defined.

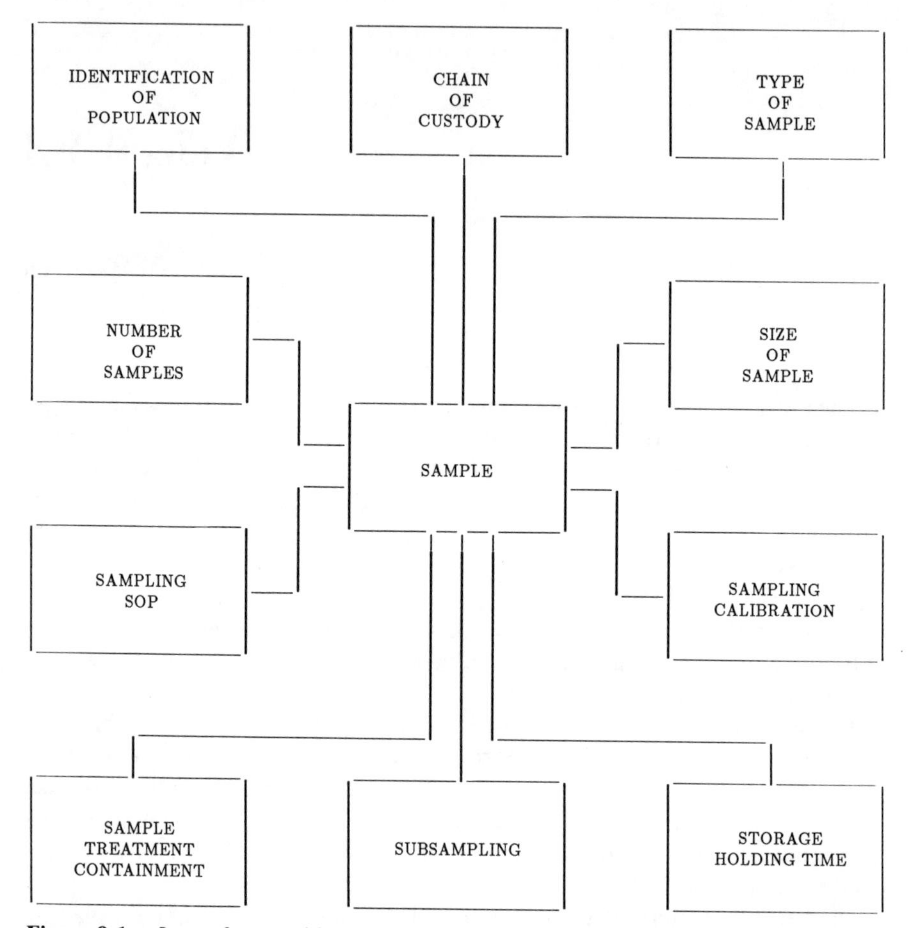

Figure 8.1. Items for consideration when developing a sampling plan.

Some Sampling Terms Defined [77]

Bulk sampling – Sampling of a material that does not consist of discrete, identifiable, constant units, but rather of arbitrary, irregular units.
Composite – A sample composed of two or more increments.
Gross sample – (Also called bulk sample, lot sample) One or more increments of material taken from a larger quantity (lot) of material for assay or record purposes.
Homogeneity – The degree to which a property or substance is uniformly distributed throughout a material. Homogeneity depends on the size of the units under consideration. Thus, an intimate mixture of two minerals may be inhomogeneous at the particulate level but relatively homogenous for 1-g portions.
Increment – An individual portion of material collected by a single operation of a sampling device, from parts of a lot separated in time or space. Increments may either be tested individually or combined (composited) and tested as a unit.
Individuals – Conceivable constituent parts of the population.

Laboratory sample – A sample, intended for testing or analysis, prepared from a gross sample or otherwise obtained. The laboratory sample must retain the composition of the gross sample. Often, reduction in particle size is necessary in the course of reducing the quantity.

Lot – A quantity of material of similar composition whose properties are under study.

Population – A generic term denoting any finite or infinite collection of individual things, objects, or events in the broadest concept; an aggregate determined by some property that distinguishes things that do and do not belong.

Reduction – The process of preparing one or more subsamples from a sample.

Sample – A portion of a population or lot. It may consist of an individual or groups of individuals.

Segment – A specifically demarked portion of a lot, either actual or hypothetical.

Strata – Segments of a lot or bulk that may vary with respect to the property under study.

Subsample – A portion taken from a sample. A laboratory sample may be a subsample of a gross sample; similarly, a test portion may be a subsample of a laboratory sample.

Test portion – (Also called specimen, test specimen, test unit, aliquot) That quantity of a material of proper size actually used for measurement of the property of interest. Test portions may be taken from the gross sample directly, but often preliminary operations such as mixing or further reduction in particle size are necessary.

The compositional makeup needs to be considered. Can the universe be considered to be homogeneous? That is to say, can every conceivable sample be said to have essentially the same composition, or is the composition expected to vary?

In the latter case, is the heterogeneity of a general nature or is there the possibility that "strata" exist that differ somewhat from each other but which may have within-strata variability, as well? In this context, strata is a generic term, denoting such distinguishing features as physically discrete layers, production lots, or ill-mixed regions, for example.

A third type of heterogeneity consists of localized areas, pockets, or even individual items scattered throughout an otherwise homogeneous matrix. In such cases, a decision may need to be made whether these are part of the universe, and hence must be considered and measured, or whether they are outliers or even foreign objects and should be excluded. One very difficult special case of the above is the possible occurrence of a single or limited number of "odd balls" in a lot which would have serious consequences if present and not detected. Cases of tampering and sabotage are examples of this. If the risk is great enough, total (100%) sampling and inspection may be required.

The question of stability of composition needs to be considered in some cases. The composition of a sample may change once it is removed from its natural matrix or environment, due to kinetic effects or interaction with a container, radiation, or air, for example. It may need to be stabilized or measured within some safe holding period.

In addition to the above, it is necessary to know:

- what substances are to be measured
- what level of precision is required
- what compositional information is needed
  - —mean composition
  - —extremes of composition
  - —variability of composition

SAMPLING PLAN

Development of a sampling plan is one of the most important parts of the overall planning process, and all of the considerations related to planning as discussed in Chapter 7 apply to this activity as well. A carefully designed sampling plan is ordinarily required to provide reliable samples. The plan should address type and number of samples to be taken, the details of collection, and the procedures to be followed.

The types of samples usually encountered are:

Representative Sample: A sample considered to be typical of the universe of concern and whose composition can be used to characterize the universe with respect to the parameter measured.

Systematic Sample: A sample taken according to a systematic plan, with the objective of investigating systematic variability of the universe of concern. Systematic effects due to time or temperature are typical matters of concern.

Random Sample: A sample selected by a random process to eliminate questions of bias in selection and/or to provide a basis for statistical interpretation of measurement data.

Composite Sample: A sample composed of two or more increments that are combined to reduce the number of individual samples needed to average compositional variability.

There are basically three kinds of sampling plans that can be used in a measurement process [142A]. Intuitive sampling plans may be defined as those based on the judgment of the sampler. General knowledge of similar materials, past experience, and present information about the universe of concern, ranging from knowledge to guesses, are used in devising such sampling plans. Because the samples are based on judgment, only judgmental conclusions can be drawn when considering the data. In the case of controversy, decisions on acceptance of conflicting conclusions may be based on the perceived relative expertise of the those responsible for sampling.

Statistical sampling plans are those based on statistical sampling of the universe of concern and ordinarily can provide the basis for probabilistic conclusions. Hypothesis testing can be involved, predictions can be made, and inferences can be drawn. Ordinarily, a relatively large number of samples will need to be measured if the significance of small apparent differences is of concern. The conclusions drawn from such samples would appear to be noncontroversial, but the validity of the statistical model used could be a matter of controversy.

Protocol sampling plans may be defined as those specified for decision purposes in a given situation. Regulations often specify the type, size, frequency, sampling period, and even location and time of sampling related to regulatory decisions. Not specifically following any part of the protocol could be reason for discrediting a sample. The protocol may be based on statistical or intuitive considerations but is indisputable once established. Testing for conformance with specifications in commercial transactions is another example. Agreement and definition of what constitutes a valid sample and the method of test are essential in many such cases.

When decisions are based on identifying relatively large differences, intuitive samples may be fully adequate. When relatively small differences are involved and the statistical significance is an issue, statistical sampling will be required.

When the number of samples required by a statistical sampling plan may be infeasible, a hybrid plan involving intuitive simplifying assumptions may be used. These assumptions should be stated explicitly whenever a statistical interpretation of the results of measurement is made. Unfortunately, this is not always done, and statistically based conclusions are sometimes made when they cannot be justified. Whatever kind of sampling plan is developed, it should be written as a protocol containing procedures (SOPs) that should be followed. It should address:

- when, where, and how to collect samples
- sampling equipment, including its maintenance and calibration
- sample containers, including cleaning, addition of stabilizers, and storage
- criteria for acceptance and/or rejection of samples
- criteria for exclusion of foreign objects
- sample treatment procedures such as drying, mixing, and handling prior to measurements
- subsampling procedures
- sample record keeping such as labelling, recording, and auxiliary information
- chain of custody requirements

Peer Review

Peer review is an important adjunct to the development of sampling plans. The more critical the analytical decision process, the greater the need for such a review. The concurrence of other experts is especially important in the intuitive planning mode. Even in the case of statistical planning, the model needs to have peer approval. As simplifying assumptions are introduced, the need for peer review increases.

In the case of controversial issues, the limitations of the analytical program, based both upon sampling and measurement, need to be considered in advance. Compromises ordinarily are necessary due to constraints of time and cost. A cost-benefit analysis that will not meet serious objections from either technical experts or any adversarial parties is a virtual necessity, or no one will be satisfied with the experimental results.

Training

In many cases, training of sampling personnel may be required. This is true especially when new sampling operations are undertaken or when new personnel are involved. In monitoring programs, all samplers should have the same training if comparable samples are to be obtained. In complex situations, dry runs could be helpful in ensuring understandability of protocols and in identifying problems that may need to be solved prior to a sampling program. Any significant deviations from a sampling plan could raise questions about and even vitiate an otherwise valid measurement program. Accordingly, problems should be anticipated as far as possible, and corrective actions should be a part of the protocol, rather than left to the judgment of the sampler.

Sample Management

Some kind of sample management is becoming imperative in modern analysis. In small programs, informal practices may be sufficient. In large and/or complex programs, a more formal approach is necessary. Consistent use of well-designed forms is the least that should be done, but data management systems may be necessary in many cases. A sample custodian may be required to ensure orderly management of samples with provision for safeguarding samples at all times.

A chain of custody that provides unbroken evidence of where a sample is located at all times may be a necessity for legal defensibility in some cases but can significantly expedite the measurement process and add confidence to the results in almost any situation.

Safety Considerations

Because environmental samples invariably involve undesirable if not hazardous materials, they must be handled with respect. Special and dedicated equipment and special facilities may be needed to prevent cross contamination of space and other samples. Special training in the use of the above may be needed.

Persons engaged in sampling hazardous sites may need initial and periodic medical examinations to ensure that they have not contracted medical problems related to the materials with which they are involved, and consideration of possible liabilities due to mishandling is recommended when developing sampling plans and strategies for hazardous materials.

Problem of Subsampling

All that is said here regarding sampling of a universe pertains as well to the laboratory sample. A universe is sampled by one of the three approaches already described to provide samples for laboratory measurement. Ordinarily, these can be much larger than the increments that are finally measured, and the selection of the latter can become a sampling problem on a microscale. Unless the laboratory sample is homogeneous, phase separations and similar problems can confound the measurement program. Some form of homogenization may be required, ranging from simple mixing to grinding or blending. Statistical sampling of laboratory samples is usually infeasible, hence, intuitive subsampling may have to be followed. The protocol for the measurement program should anticipate subsampling problems and address this subject. Otherwise, significant variability could result from variation in the intuitive approach by laboratories, individuals, or even the same individual on various occasions.

CRITICAL MATTERS RELATED TO SAMPLES

Critical matters related to samples can be both qualitative and quantitative. In the first category, what is sampled is of prime importance. There should be no questions about identification; however, the appropriateness of the sample may be questioned. This involves the appropriateness of the model for the measurement problem but also of the model for sampling.

The integrity of the actual sample measured is related to questions of preservation involving containment, storage, and holding time, possible dilution or contamination, and subsampling. Questions of substitution, mislabeling, and accidental confusion may also need to be addressed.

Quantitative questions relate to calibrations of sampling equipment and ancillary data as they are important. Loss of samples for any reason can lead to incomplete data sets that can perturb the best designed sampling programs. Many of the above problems can be alleviated by good documentation practices.

QUALITY ASSURANCE OF SAMPLING

The sampling operations must be supplemented by a credible quality assurance program embodying the two aspects of quality control and quality assessment. When data quality goals are established, the quality control necessary to achieve them can be planned. Sources of excessive random and systematic error need to be identified, and protocols must be developed to reduce them to acceptable levels. Sample takers must be trained to follow them precisely.

Quality assessment of the sampling operation is restricted to monitoring for contamination, using appropriate blanks, and auditing the documentation. Because the degree of adherence to protocols is one of the objectives of an audit, the importance of the content of the protocols is obvious.

SAMPLE UNCERTAINTIES

The uncertainties in the data due to the sample always need to be evaluated. The variance of the population can be estimated easily, provided the samples are randomly selected and amenable to statistical analysis. Bias of samples with respect to a population can result from a defective model, but this may be difficult to establish. Some discrimination is often built into a model. For example, the component of interest may be considered to be in a specific particle size fraction, and screening may be used to separate it from an undesirable fraction. If this is inappropriate, a bias could result.

The sampling operation can introduce both systematic and random errors. Calibration errors can cause problems of the first type, while variability of operations such as sieving or extraction are examples of the latter kind. It may be impossible to quantify the individual components of sampling variance. However, the overall sampling variance can be evaluated by taking a number (at least 7) of samples under conditions where the samples are expected to be essentially identical. The total variance consists of the sum of that due to the samples and to their measurement. Thus:

$$s^2_{total} = s^2_{sample} + s^2_{measurement}$$

The measurement variance is subtracted from the total variance to obtain sample variance.

Of course, it is the variance of the sample population that is of most concern when answering questions by measurement. The variance of the samples measured is related to that of the population and sampling by the expression:

$$s^2_{sample} = s^2_{population} + s^2_{sampling}$$

Obviously, it is good advice to take all the care necessary to make sampling variance negligible.

Some measurement programs require ancillary data such as pressure, temperature, flow, and moisture content, to reduce the data to standard conditions. Both random and systematic errors in measurement of these parameters can reflect similar sources of error in the measurement data.

Some samples could be affected by deterioration during collection, transit, or storage. Interactions with other constituents, container walls, and transfer lines are other sources of uncertainties. While these should be addressed in sampling plans, the bounds for bias resulting from these sources should be considered.

Stratification

Stratification, when possible, is an insidious source of error in analytical samples. Samples that were initially well-mixed may separate, partially or fully, over a period of time. It may be difficult (perhaps impossible) to reconstitute them, and even the need to do so may not be apparent. Examples include a mixture of solids that may separate due to gravitational and/or vibrational forces, emulsions that could demulsify, and water samples containing suspended matter that later could plate out on the container walls.

Whenever stratification is possible, care should be exercised to reconstitute the sample, to the extent possible, each time a subsample is withdrawn. Otherwise, problems caused by poor mixing can become even more serious as the ratio of sample increment to residual sample increases. Any apparent uncompensated uncertainties resulting from segregation in its various aspects should be considered when evaluating measurement data.

Holding Time

When degradation is possible, samples should be measured before any significant change has occurred. This leads to the concept of "holding time," defined as the maximum period of time that can elapse from sampling to measurement before significant deterioration can be expected to occur.

A simple statistical procedure to evaluate holding time is illustrated in Figure 8.2 together with graphical instructions for its evaluation. In brief, a sufficiently large quantity of a typical sample, is held under a condition of storage and/or preservation for a period of time. Increments are withdrawn at the start and at intervals and measured in duplicate. The differences of the duplicate measurements allow estimation of the standard deviation of measurement, s, as discussed in Chapter 4. The smallest difference in two measured values that is significant at the 95% level of confidence is $2\sqrt{2}$ s, which is approximately equal to 3s.

The means of the duplicate measurements are plotted with respect to time

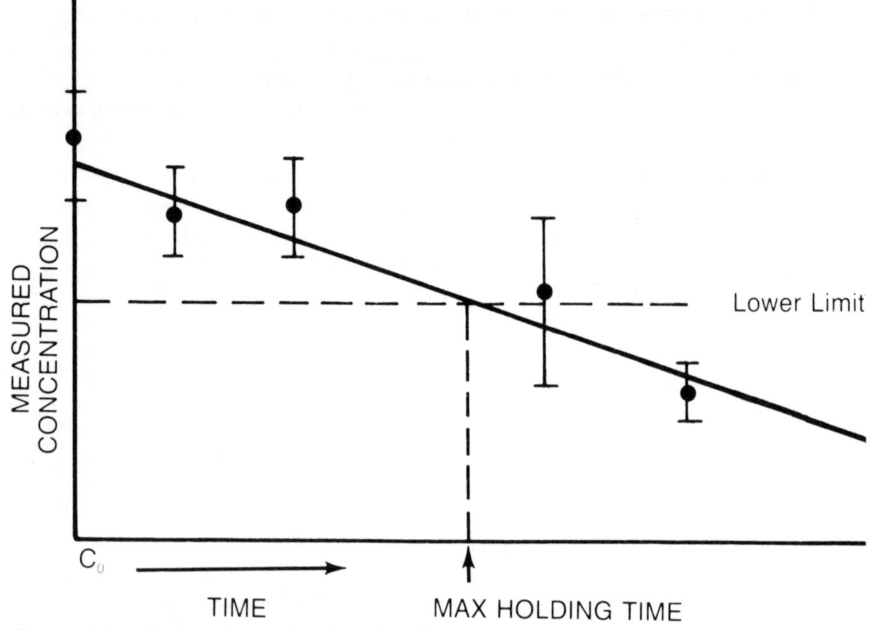

Figure 8.2. Estimation of holding time.

and graphically fitted with a straight line, which is good enough for the intended purpose. The line labelled lower limit is the difference of the extrapolated original concentration, C_0, minus 3s. The point of intersection of the lower limit and the fitted line represents the maximum holding time before any significant deterioration has occurred.

If the holding time is considered to be inconveniently short, the conditions of storage and/or stabilization must be modified necessitating a new determination of holding time. Very transitory samples may need to be measured immediately after sampling.

STATISTICAL CONSIDERATIONS FOR SAMPLING

Measurement Situations

The various kinds of measurement situations encountered by analytical chemists are outlined in Table 8.1. Situation A is the most desirable but is seldom encountered. In this case, neither the measurement nor the sample variance is significant, hence a single measurement on a single sample could provide all of the information required to make a decision. Only the measurement variance is significant in situation B. This can be known in advance when the measurement system is in statistical control, hence a single measurement of

Table 8.1. Measurement Situations

	Significant	Not Significant
A. Measurement Variance		X
Sample Variance		X
B. Measurement Variance	X	
Sample Variance		X
C. Measurement Variance		X
Sample Variance	X	
D. Measurement Variance	X	
Sample Variance	X	

a representative sample can be used for decision purposes. Situation C is more complex in that more than one sample measurement will be needed in order to relate the data to the universe of interest. However only single measurements of the samples would need to be made. Ordinarily, the sample variance cannot be known with sufficient exactitude prior to the measurements (although there may be a more or less reliable estimate of it) so that a sampling program will need to be developed and followed to evaluate it.

Situation D occurs frequently, especially in environmental analysis. A sampling plan, involving multiple samples, will be needed as in C. Unless the measurement variance is known and documented, multiple measurements may need to be made as well, and a statistical analysis of variance of the results will be necessary in order to evaluate the several variances.

Obviously, knowledge, or at least estimates of the several variances, will be needed to design the sampling plan, the measurement program, and to interpret the measurement data.

Statistical Sampling Plan

Statistics can provide several kinds of information in the development of sampling plans and the evaluation of measurement data. These include:

- the limits of confidence for the measured value of the population mean
- the tolerance interval for a given percentage of the individuals in the population
- the minimum number of samples required to establish the above intervals with a selected confidence level

There are several conditions that must be met when making such statistical estimates. First, the samples must be randomly selected from the population of concern. Each sample must be independent of any other sample in the group. The type of distribution of the samples must be known, in order to apply the correct statistical model.

The first two conditions are realized by a sampling operation that employs a randomization process and assurance that one sample does not influence another, such as by cross-contamination.

The question of the distribution may be more difficult to answer. A Gaussian distribution is often applicable, or assumed, and such statistics are easiest to apply. However, it may be infeasible to confirm experimentally whether such a distribution (or any other) is valid because of the large number of samples that would be required.

The following discussion is based on the assumption that a Gaussian distribution applies.

Basic Assumptions

- For statistical planning, an estimated or assumed standard deviation is treated as if it were the population standard deviation.
- For statistical calculations, the standard deviation estimate, s, based on measurement data, is used in the conventional manner.
- In the analysis of variance from several sources, it is assumed that the total variance is equal to the sum of the various components:

$$s^2_0 = s^2_1 + s^2_2 + \ldots + s^2_n$$

In sampling, we will use the notation

σ_A = standard deviation of measurement
σ_s = standard deviation of samples (within a stratum)
σ_B = standard deviation between strata

Guidance in Sampling

An excellent discussion of the statistical basis for sampling is given by Provost [111] who presents nomographs to assist in the solution of some of the more complex equations that follow.

The following equations describe the minimum number of samples and/or measurements necessary to limit the total uncertainty, to a value E, with a stated level of confidence as indicated by the value of z that is selected. For the 95% level of confidence, $z = 1.96 \approx 2$.

a. Minimum Number of Measurements, n_A (σ_s negligible)

(Measurement Situation B)

$$n_A = \left(\frac{z\sigma_A}{E_A}\right)^2$$

b. Minimum Number of Samples n_s (σ_A negligible)

(Measurement Situation C)

$$n_s = \left(\frac{z\sigma_s}{E_s} \right)^2$$

In the above calculations, if n_A is more than is considered to be feasible:

- improve the precision of the methodology to decrease σ_A, or
- use a more precise method of measurement if available (smaller σ_A), or
- accept a larger uncertainty

If more samples are required than is feasible:

- use a larger sample (smaller σ_s), or
- use composites (smaller σ_s), or
- accept a larger uncertainty

c. When σ_A and σ_s Both are Significant

(Measurement Situation D)

$$E_{Total} = \left(\frac{\sigma^2_s}{n_n} + \frac{\sigma^2_A}{n_s n_A} \right)^{1/2}$$

where n_A is the number of measurements per sample.

The equation above has no unique solution in that several values of n_s and n_A will produce the same value of E. Compromises will be necessary, taking into consideration the costs of sampling and of measurement. The nomographs of Provost may be helpful in making the various estimates. Again, ways to decrease σ_s and/or σ_A will decrease the number of samples and/or measurements required.

* When Samples Come from Several Strata

$$E_{Total} = \left(\frac{\sigma^2_B}{n_B} + \frac{\sigma^2_s}{n_B n_s} + \frac{\sigma^2_A}{n_B n_s n_A} \right)^{1/2}$$

where n_B = number of strata sampled
n_s = number of samples per strata

Again, no unique solution is possible so that compromises will be required.

Cost Considerations

Usually, it is necessary to design a sampling plan with cost considerations in mind.

Let n_s = total number of samples
n_A = number of measurements per sample
C_s = cost per sample of sampling
C_A = cost per measurement
C = total cost of program
$C = n_sC_s + n_sn_AC_A$

It can be shown that

$$n_A = \left(\frac{C_s\sigma^2{}_A}{C_A\sigma_{2s}}\right)^{1/2}$$

and

$$n_s = \frac{Z^2(\sigma^2{}_s + \sigma^2{}_A/n_A)}{E^2}$$

and

$$E = \left[\frac{Z^2(\sigma^2{}_s + \sigma^2{}_A/n_A)}{n_s}\right]^{1/2}$$

If the maximum allowable (or budgeted) cost is C_m, n_A is calculated from the equation given above and substituted in the expression below to obtain n_s:

$$n_s \geq \frac{C_m}{C_s + n_AC_A}$$

The values for n_s and n_A are then substituted in the equation given earlier to calculate E, which could be larger than the acceptable limit of error. In this case, compromise will again be required.

Minimum Size of Increments in a Well-Mixed Sample

In the case of heterogeneous solids, it is well known that variability between increments increases as their size decreases. Ingamells [66] has shown that the following relation is valid in many situations.

$$WR^2 = K_s$$

where W = weight of sample analyzed, g
R = relative standard deviation of sample composition, %
K_s = sampling constant, required to limit the sampling uncertainty to 1% with 68% confidence

Once K_s is evaluated for a given sample population, the minimum weight, W, required for a maximum relative standard deviation of R percent can be calculated. If the material sampled is well mixed, the relationship holds very well. For segregated or stratified materials, the calculated value for K_s increases as W increases. This is one way that the degree of mixing can be judged.

Size of Sample in Segregated Materials

In the case of segregated materials, Visman [144] has shown that the variance of sampling, s^2_s, may be expressed by the relation

$$s^2_s = \frac{A}{Wn} + \frac{B}{n}$$

where A = random component of sampling variance
B = segregation component
W = weight of individual increment, g
n = number of increments

The error due to random distribution may be reduced by increasing W and/or n. Doubling the number of increments has the same effect as doubling the weight of each increment. However, the error due to segregation is reduced only by increasing the number of increments sampled.

The constants A and B must be evaluated experimentally. One way to do this is to collect a set of large increments, each of weight W_L, and a set of small increments, each of weight W_S. Each set is measured and the respective standard deviations, S_L and S_S, are calculated. Then

$$A = \frac{W_L W_S}{W_L - W_S} (s^2_S - s^2_L)$$

and

$$B = s^2_L - \frac{A}{W_L}$$

Visman calculates the degree of segregation

$$Z_s = \frac{B}{A}$$

and says that values for $Z_s > 0.05$ can cause serious errors in the estimation of s_s if the single sampling constant approach is used.

ACCEPTANCE TESTING

Acceptance testing is based on judging whether a lot of material meets preestablished specifications. One type involves decisions on whether a material tested meets a compositional specification such as the carbon content of a cast iron. The other consists in the extent or fraction of defective (out-of-specification) items in a lot.

If a representative sample is to be used as the basis for decision, it must be defined in the plan. If multiple samples are to be used, then the number, size, and method of collection need to be specified. The permissible variability of a material (homogeneity) is often important and decisions on conformance require statistical sampling. As the limits are narrowed, the number of samples required for decision will increase. The statistical procedures to calculate confidence intervals for means and statistical tolerance intervals, and to judge the significance of the differences of measured and specified values, are essentially the same as those discussed in Chapter 4.

The determination of whether the observed percentage of defectives is significant and exceeds some specified value requires statistical sampling of the material in question. In a single sampling plan a predetermined number of units (sample) is tested, and the number that fail the test is determined. If this exceeds a critical number, the lot is rejected, based on the probability that the entire lot has an excessive number of defective items. Multiple sampling plans involve testing a second and even a third set, in some instances, when the first lot fails the test. The maximum permissible number of defectives in the second lot (sometimes zero) is specified in order to accept the lot. General principles of this kind of testing are discussed by Besterfield [26] and by Natrella [100]. Table 8.2 is an abridgement of a table found in MIL STD-105D [96]. Based on a single sampling, the table specifies the number of samples (n) to be examined and the number of defective items that would require rejection (RE) or permit acceptance (AC) on the belief, with 90% confidence, that the lot does (rejection) or does not (acceptance) contain more than the given percentage of defective items (acceptable quality level).

An interesting special case of the above is the decision on whether there are any defective items in a lot. One can never be sure that this is so unless every item is tested (100% inspection). If only one item escapes inspection, it is

Table 8.2. MIL STD-105D Single Normal Sampling Plans

	Acceptance Quality Level								
	2.5%			4.0%			6.5%		
Lot Size	n	Ac	Re	n	Ac	Re	n	Ac	Re
2–15	5	0	1	3	0	1	2	0	1
16–25	5	0	1	3	0	1	8	1	2
26–50	5	0	1	13	1	2	8	1	2
51–90	20	1	2	13	1	2	13	2	3
91–150	20	1	2	20	2	3	20	3	4

Table 8.3. Defective Items in a Lot

Number of Samples Tested but No Defectives Found	Percentage of Items That Could Be Defective At Indicated Probability		
	10 %	5 %	1 %
1	68.4	77.6	90
2	53.6	63.2	75.5
4	43.8	52.7	68.4
6	31.9	39.3	53.6
8	25.0	31.2	43.8
10	20.6	25.9	36.9
12	17.5	22.1	31.9
14	15.2	19.3	28.0
16	13.4	17.1	25.0
18	12.0	15.3	22.6
20	10.9	13.9	20.6
30	7.4	9.5	14.2
50	5.1	8.0	11.0
100	2.6	4.0	6.0
250	1.0	2.0	2.5
1000	>1	>1	1

possible that it is defective. As the number of items inspected and found to be acceptable increases, the risk of concluding that the entire lot is acceptable diminishes and can be statistically evaluated.

Natrella [100] gives tables (A-22, A-23, A-24) that define the percentage of a lot that could be defective, with a given probability, when no defects have been observed in any of the samples tested. These tables are based on the assumption that the number of items tested is small when compared with the total number of items in the lot. Table 8.3 is compiled from the tables cited above.

Table 8.4. Quality Level Inference when No Defects are Observed

Number of Tests Without Failure vs. Quality Level and Confidence

Minimum Percentage Acceptable	Possible Defectives, ppm	Number of Tests Without Failure Confidence, percent		
		90	95.0	99
99.9999	1	2302590	2995730	4667150
99.999	10	230259	299573	460517
99.99	100	23026	29957	46052
99.9	1000	2303	2996	4605
99.	10,000	229	298	459
95.	50,000	45	58	90
90.	100,000	22	29	44
80.	200.000	11	14	21
70.	300,000	7	9	13
60.	400,000	5	6	9
50.	500,000	4	5	7

Excerpted from "Quality", Hitchcock Publishing Co., Wheaton Il, March, 1986, p75. Used with permission.

Schilling [122] describes a sampling plan for lot acceptance when the number of items tested is a significant fraction of the size of the lot.

Table 8.4 is presented to further illustrate the quality level that could be inferred with a stated confidence when no defectives are found on examination of a given number of samples. For example, when 9 samples are examined and all are found to be acceptable, one could conclude with 95% confidence that at least 70% of the lot is acceptable or with 99% confidence that at least 60% is acceptable. To support confidence that most of a lot is acceptable requires a much larger number of samples, none of which can be found to be defective. Thus 298 such samples would support a 95% confidence that at least 99% of a lot was not defective. Even if 4,665,170 samples were tested without failure, there is a 1% chance (confidence level 99%) that there could be one defective item in a million items (1 ppm).

MATRIX-RELATED SAMPLING PROBLEMS

Ways to sample specific kinds of materials will be found in the scientific literature. Many ASTM standards deal specifically with such problems. Sampling equipment has been developed and guidance documents have been written for specific situations. Virtually all of the recent significant sampling literature is reviewed in Kratochvil, et al. [79]. The following sections discuss sampling problems from a matrix basis and from a general point of view.

Sampling Gases

Gases are considered to be microhomogeneous, hence the number of samples required is related largely to differences due to time and location of sampling. Gases in the atmosphere can be stratified due to emissions from point sources and to differences in density of components. Mixing by diffusion, convection, and mechanical stirring can be more inefficient than might be imagined. When gas blends are prepared in cylinders, it can take considerable time for the mixtures to equilibrate for the same reasons.

The techniques of gas sampling are described in a number of references [79]. Personal monitors located in the breathing zone of workers are used to evaluate the exposure of workers where stratified levels can occur in the work place atmosphere due to the nature of industrial operations [79].

Environmental gas analysis often involves real-time measurement of samples extracted from the atmosphere. The siting of monitoring stations is a critical aspect of such measurements. The possible degradation of samples during transit in sampling lines also needs to be considered.

Sampling may consist of absorbing some component of interest from an atmosphere in order to concentrate it to measurable levels or for convenience of later measurement. Efficiencies of absorption and desorption, and uncertainties of auxiliary data such as temperature, pressure, and volume sampled need to be considered when interpreting such measurement data.

Gas sampling can consist of collecting a portion in a container. The possible interaction of components with the walls of the container and also the possible deterioration and/or interaction of constituents need to be considered.

Grab samples of gases may be collected by opening a valve of a previously evacuated container. Composite samples may be obtained by intermittently opening the valve for predetermined intervals, selected to obtain equal increments of gas, based on pressure drop considerations.

Gas samples are sometimes collected by condensation in a refrigerated trap. One must remember that the composition of gas evolved on warming can vary due to differential evaporation unless the entire sample is evaporated before withdrawal of any part of it. The same problem can be encountered in withdrawal of the contents of a cylinder gas containing easily condensed components.

Sampling Liquids

What has been said about gases applies in principle to the sampling of liquids. Stability of a collected sample and interaction with containers is of major concern in some cases. Ways to stabilize samples have been developed, and environmental regulations often specify what must be done in this regard. The concept of holding time (see page 63) is especially applicable to liquids and especially to aqueous samples.

Liquid samples are often large, to provide the possibility of multiparameter measurement. Methods of transport need to be considered to minimize breakage and loss of contents. Samples may appear to be homogeneous, but phase separations can occur as a result of temperature changes, freezing, or long standing. When such occur, it may be impossible to fully reconstitute the sample.

Sampling Solids

The methods used to obtain solid samples depend on whether the material is massive, consists of aggregates, or is fine granular. Heterogeneity is a common characteristic, and the statistical sampling considerations already discussed pertain especially to such samples. Questions of stability ordinarily are not of major concern, but air oxidation and/or moisture interactions can cause problems, especially for finely ground materials with large surface areas.

Subsampling of solids can be a major problem, as has already been mentioned.

Sampling Multiphases

All of the possible combinations—liquid-solid, liquid-gas, solid-gas, liquid-liquid, solid-solid, and liquid-solid-gas—can be encountered. The sampling plan applicable will depend on whether a specific analyte, a single phase, a complete analysis, or any variation of the above is of concern.

Once a sample has been obtained and/or removed from its normal environment, the possibility of phase changes and disruption of phase equilibria must be considered and addressed as necessary. There can be changes due to differential volatility, absorption, settling of suspended matter, demulsification, and related physical and chemical effects that could change a sample drastically.

In some cases, the phase of interest is separated from its matrix or carrier phase prior to measurement. The removal process must be efficient. The problem of isokinetic sampling of particulates in emission sources is an important example. Should the removal process discriminate between components of a sample, the resulting physical bias could bias the analytical results as well.

Because phase separations can occur in some cases, ways to prevent or minimize them should be sought. In some situations, it may be virtually impossible to reconstitute the original sample once a phase separation has taken place. Even if no loss has occurred, the analysis could be unduly complicated. Failure to recognize phase changes could produce significant analytical errors.

CHAPTER 9

Principles of Measurement

Measurements usually are made because a quantitative value for some substance or constituent thereof is needed for some purpose. Ordinarily, one speaks of measurement or determination of the parameter of interest, but this is seldom the case. Rather, one must be content with measuring some property for which there is an instrument response that can be related to the concentration or amount of the substance of interest in a sample of the material or in some other material (e.g., a solution) prepared from the sample. Furthermore, a suitable system should be available to make the desired measurement, and the system should be maintained in a state of statistical control throughout the measurement process [46]. A further requirement is that the measurement system can be and is calibrated with respect to the substance of interest.

TERMINOLOGY

The nomenclature of methodology often is loosely used, so a short discussion of this topic may be in order at this point. The hierarchy of methodology, proceeding from the general to the specific, may be considered as technique → method → procedure → protocol. These terms may be defined as follows:

Technique – A scientific principle that has been found useful for providing compositional information.

Method – An adaptation of a technique to a specific measurement problem.

Procedure – The written directions considered to be necessary to utilize a method.

Protocol – A set of definitive instructions that must be followed, without exception, if the analytical results are to be accepted for a specific purpose.

In addition to these, several other terms are used when describing methodology or measurement data. Some of them are:

Absolute Method	– Method in which characterization is based entirely on physical (absolute) standards.
Comparative Method	– Method in which characterization is based on chemical standards (i.e., comparison with such standards).
Reference Method	– A method of known and demonstrated accuracy.
Standard Method	– A method of known and demonstrated precision issued by an organization generally recognized as competent to do so.
Standard Reference Method	– A standard method of demonstrated accuracy.
Definitive Method	– A method of known accuracy that has been accepted to establish the property of some material (e.g., reference material) and/or to evaluate the accuracy of other methodology (term widely used in clinical analysis).
Routine Method	– Method used in routine measurement of a measurand. It must be qualified by other adjectives since the degree of reliability is not implied.
Field Method	– Method applicable to nonlaboratory situations.
Trace Method	– Method applicable to ppm range.
Ultra Trace Method	– Method applicable below trace levels.
Macro Method	– Method requiring more than milligram amounts of sample.
Micro Method	– Method requiring milligram or smaller amounts of sample.

The terms above are operational descriptors of methodology but do not define its quality or usefulness for a specific purpose. Various attempts have been made to classify methodology on the basis of achievable precision and accuracy, but no consensus exists on this respect. Table 9.1 represents a classi-

Table 9.1. Classes of Methods

Class	Precision/Accuracy (P/A)	Nomenclature
A	<0.01%	Highest P/A
B	0.01–0.1%	High P/A
C	0.1–1%	Intermediate P/A
D	1–10%	Low P/A
E	10–35%	Semiquantitative
F	>35%	Qualitative

For trace analysis, move classes and corresponding nomenclature down one step.
For ultra-trace analysis, move classes and corresponding nomenclature down two steps.

fication which has been found to be useful when considering the relative performance characteristics of candidate methodology.

MEASUREMENT IS A COMPARISON PROCESS

Measurement is basically a comparison of an unknown with a known. In some cases, the comparison is direct, as in the determination of mass using an equal-arm balance. In other cases, it is indirect, such as when using a spring balance. Direct chemical measurements consist of comparisons on a real time basis or intermittent alternations of standard and unknown. In indirect measurement, the scale readout of an instrument may be calibrated at intervals of time which should be selected so that no significant changes of the scale factor occur during that period of time. Specific details of calibration are discussed in Chapter 10.

PRINCIPLES OF CHEMICAL MEASUREMENT

Except for the assay of pure substances, the constituent(s) of interest is usually contained in a matrix that may or may not influence its measurement. Some measurements can be made in the native matrix with little or no modification thereof. Emission spectroscopy is an example of this. In others, a matrix modifier may be added to facilitate the measurement or to provide a matrix common to a variety of measurements and thus minimize calibration problems. Removal of the constituent of interest from its matrix is a common practice to eliminate matrix problems, to concentrate the constituent, to enhance detection limits, or to increase measurement accuracy. The removal process may utilize dissolution of a sample in a suitable solvent (or reactant), distillation, extraction, filtration, or chromatographic separation, for example. Whenever a removal operation is a part of the measurement process, the criticality of the steps that are involved needs to be understood thoroughly, and appropriate tolerances must be established and maintained if accurate results are to be obtained. It should be remembered that small modifications of the removal procedure as well as the final measurement process can constitute a modified if not a different method of measurement with corresponding changes in the accuracy of the output. To minimize unintentional changes, the development and utilization of an optimized standard operations procedure (SOP) is highly recommended.

Table 9.2. Figures of Merit

Figures of Merit
Essential Characteristics
Precision
Detection level
Sensitivity
Bias
Selectivity
Useful range
Desirable Characteristics
Speed
Low Cost
Ruggedness
Ease of operation

FIGURES OF MERIT

The important operating characteristics of a method are sometimes called figures of merit since they provide the basis for selection of a methodology that is appropriate for an intended purpose. Figures of merit are not constants of methodology since their numerical values are influenced by the user and thus will depend on the way the methods are used. Accordingly, typical values are all that are ordinarily available, but these are adequate in most cases. Figures of merit are conventionally based on the interpretation of a single measurement, and some of them can be reduced by the factor $1/\sqrt{n}$ if n independent replicate measurements are possible.

The first six of the figures of merit listed in Table 9.2 are critical for selection of methodology and need to be evaluated quantitatively. Unless there is a reasonable match with the requirements for the data, a methodology is essentially useless.

Precision

The precision, detection level, and sensitivity are easily evaluated from knowledge of the standard deviation as a function of measurement level. Estimates based on at least 6 degrees of freedom should be obtained at each point evaluated to provide useful statistics. If such information is not available, the analyst may need to make the measurements required to provide it. The standard deviations estimated from measurements made at a minimum of 3 levels (low, mid-range, and high range) may be plotted as a function of level as shown in Figure 9.1, and the simplest relation is graphically fitted to the data (usually a straight line).

The standard deviation at any concentration level represents the expected precision of measurement at that point. The value for s_o, the value of the standard deviation as $C \rightarrow 0$, obtained by extrapolation, is used to evaluate the limit of detection.

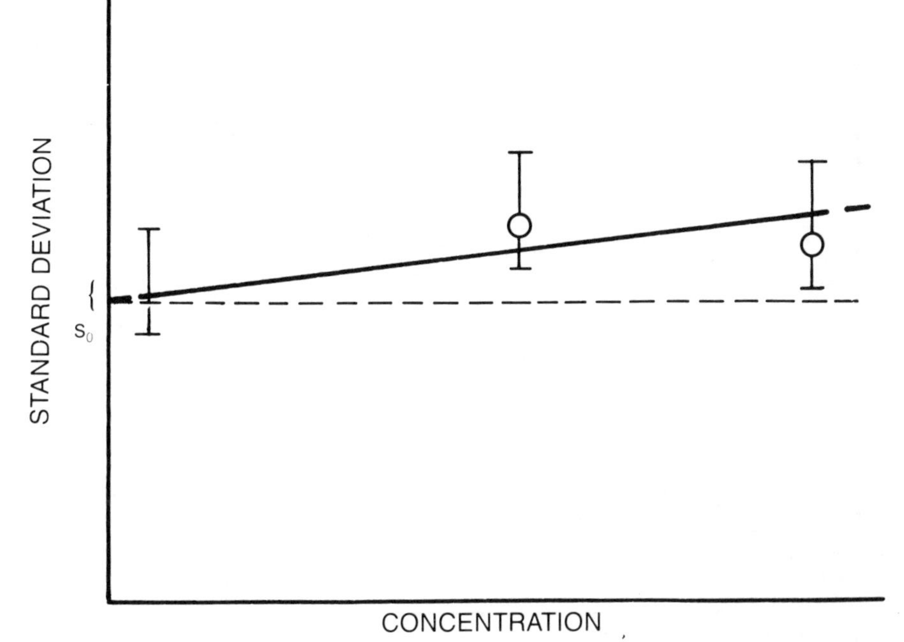

Figure 9.1. Basic statistics for a method of measurement.

Sensitivity

The sensitivity is evaluated from the value of s_c, the standard deviation at the concentration level of interest, and is equal to $S_c = \Delta C = 2\sqrt{2}\ s_c$ which represents the minimum difference in two samples of concentration $\approx$ C, that can be distinguished with a 95% confidence.

Limit of Detection and Limit of Quantitation

A measured value becomes believable when it is larger than the uncertainty associated with it. The point at which this occurs is called the limit of detection (LOD) and is defined arbitrarily as $3s_o$, as discussed later [2,3]. The lower level where measurements become quantitatively meaningful has been called the limit of quantitation (LOQ) and is defined arbitrarily as $LOQ = 10\ s_o$ [2,3]. At this concentration, the relative confidence in the measured value is about ± 30% at the 95% probability level. The upper level of reliable measurement, for all practical purposes, is at the limit of linearity (LOL). The useful range of the methodology is thus that range of concentrations between the LOQ and the LOL (see Figure 9.2).

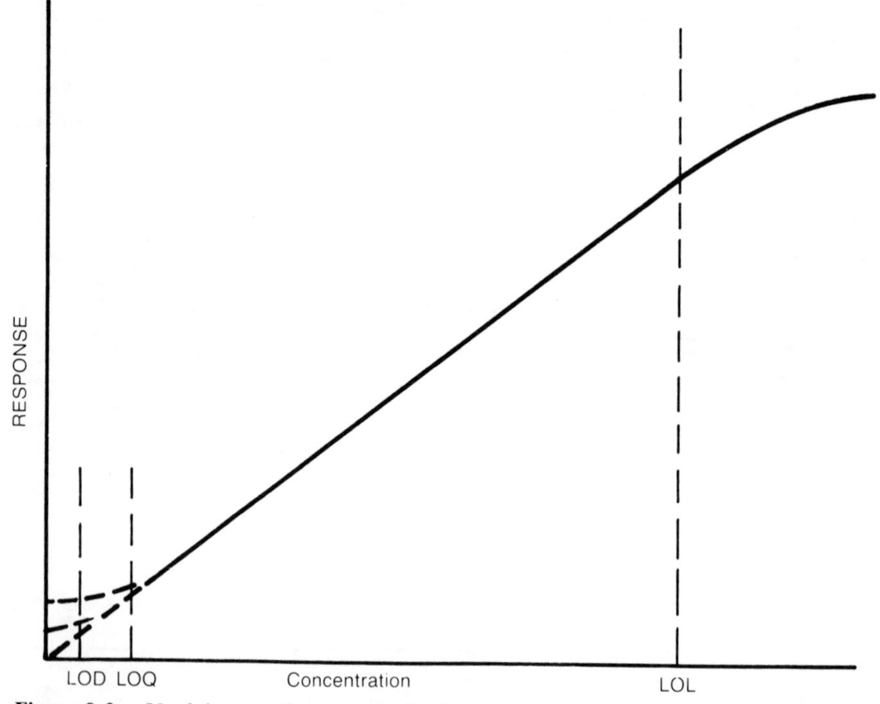

Figure 9.2. Useful range for a method of measurement.

In ultratrace analysis, the region of concern is often close to the LOQ and/or the LOD. In this case, the value of s_o is all that is required to characterize the methodology and the precision of the results obtained. While s_o is best evaluated by the extrapolation method as shown in Figure 9.1, it can be estimated with sufficient confidence by repetitive measurements of a single sample (or a reasonable number of duplicates) containing measurable concentrations of the analyte, at a concentration level close to the LOQ. Samples with concentrations of approximately $20s_o$ are recommended for this purpose.

The term "limit of detection" is used in several different situations which should be clearly distinguished. One of these is the instrument limit of detection (IDL)—or even that of a detector within an instrument—and this is based on the ability to distinguish between "signal" and "noise." If the method used is entirely instrumental, i.e., no chemical or other procedural steps are involved, the signal-to-noise ratio has some relation to the LOD. However, the statistical significance is often vague. Whenever other steps are involved in the measurement, such considerations relate only to the measurement step, the variance of which may be small with respect to that from other sources of variance in the methodology. Thus, improvement of signal-to-noise ratio by instrumental modifications, for example, would not necessarily lower the LOD.

The method limit of detection (MDL) [55] is based on a method's ability to determine an analyte in a sample matrix, regardless of its source of origin. This is based on the variance for the analysis of samples. A sufficient number of samples at each of 3 levels (at least 7) are analyzed, and the standard deviations of the results are estimated. These are plotted as shown in Figure 9.1 to obtain s_0 by extrapolation. The MDL at 95% confidence is defined as $3s_0$. Alternatively, a single sample containing a small but measurable amount of analyte may be measured at least 7 times to evaluate s. If the concentration level is about twice the LOQ, the value obtained for s may be considered to be equivalent to s_0 without significant error. Otherwise, the value for s_0 should be obtained by extrapolation as discussed above. When obtaining s_0 by extrapolation, one must be sure that the lowest value of analyte used is reasonably close to "zero," otherwise the extrapolation distance may be excessive. The lowest level may be obtained by iteration in the case of completely unknown methodology.

Then, there is the case of detection of a level of concentration above that of a blank, a natural background, or a control site—a decision process of considerable importance in some monitoring programs. In such cases, the question of how much of an analyte might have been present but not detected is sometimes raised. This is related to, but not the same as, the above-mentioned LODs.

Figure 9.3 is presented to clarify the concepts of LOD and LOQ of a measurement process. In the figure, the relative uncertainty of a measured value, expressed as a percentage, is plotted with respect to the concentration of the analyte, expressed as multiples, N, of the standard deviation. At the 95% level of confidence,

$$\text{Relative uncertainty} = 2\sqrt{2}\,\frac{s_0}{X} \times 100$$

$$= 2\sqrt{2}\,\frac{s_0}{Ns_0} \times 100 = 2\sqrt{2}\,\frac{1}{N} \times 100$$

From the above relation, it can be seen that the relative uncertainty is $\pm$ 100% for an analyte concentration of $3s_o$. This is the limiting level where the measured value exceeds its uncertainty and is thus the limit of detection, LOD, with 95% confidence. If a higher confidence is required, the LOD would be correspondingly larger.

As the concentration level increases beyond this point, the relative uncertainty decreases so that semiquantitative data is obtainable. At some higher level, the relative uncertainty will have decreased to a point where the data attain quantitative significance. The American Chemical Society Committee on Environmental Improvement [2,3] calls this the limit of quantitation and arbitrarily defines it as $10s_o$, corresponding to a relative uncertainty of about $\pm$ 30%.

In considering questions of detection (and also IDL and MDL), it should be

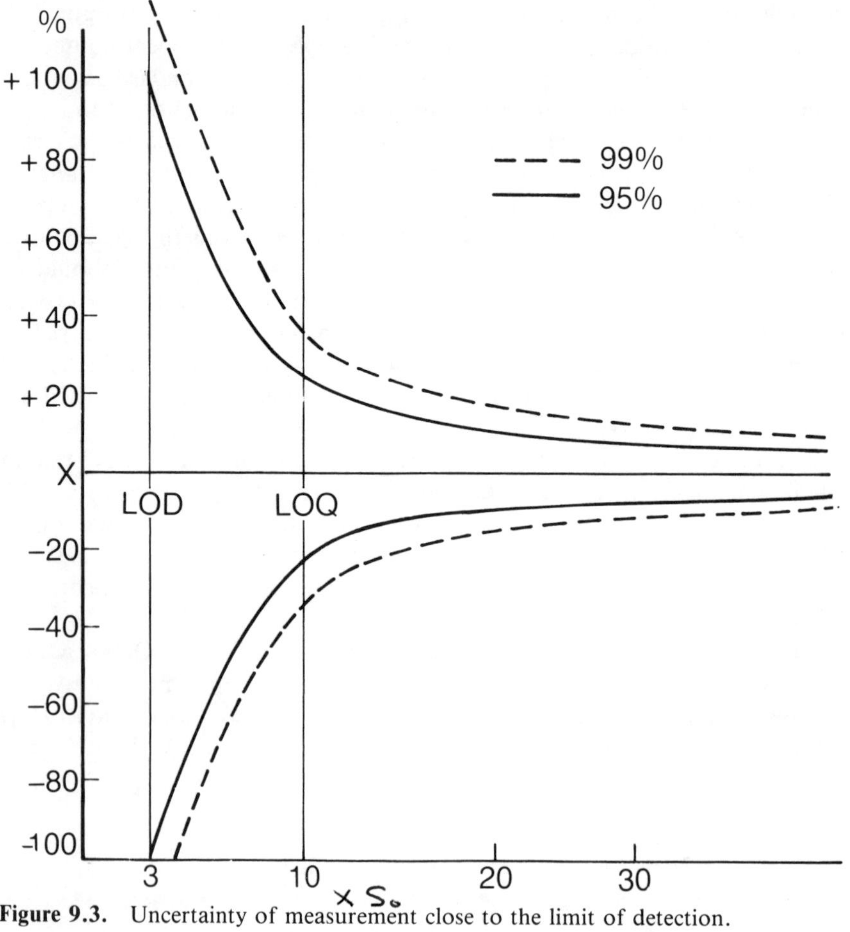

Figure 9.3. Uncertainty of measurement close to the limit of detection.

remembered that the standard deviations involved are not unique constants of the methodology; ordinarily they will depend on the expertise of the analyst, the quality control procedures utilized, and even on the matrix measured. Thus two analysts or laboratories using the same methodology can show significant differences in precision, and hence their LODs will differ.

In Figure 9. 1, the error bars indicate the confidence limits for the estimates of the standard deviation, s. These limits are asymmetrical and are large when s is based on a relatively small number of degrees of freedom. When questions of detection are of special concern, it is recommended that s_0 should be reestimated regularly and the values control charted. If this is done, a simple range control chart (see Chapter 14) for a reasonable number of duplicate samples of analyte level reasonably close to the LOQ can attest to maintenance of the method detection limit (MDL) [2,3,55]. Otherwise, at least 7 replicate measurements should be made each time the MDL is reevaluated.

Short-Term and Long-Term Standard Deviation

Whenever the variability of a measurement process is of concern (as measured by a standard deviation), it should be remembered that often this is the result of several simultaneously acting sources of variability, some of which may be invariant over short periods of time. For example, the same calibration may be used for a number of measurements, hence the calibration variance is nonexistent for all of them. Also, instrumental operational and laboratory environmental fluctuations may be relatively constant over short periods of time. Accordingly, short-term standard deviations are usually smaller than long-term standard deviations. That is to say, the measurement system is more precise over short intervals than long intervals of time. The same concept applies to detection levels, limits of quantitation, and sensitivity. The short-term standard deviation provides guidance on how many repetitive measurements should be made at any one time to attain a desired level of precision. This gives rise to the term repeatability. The long-term standard deviation is a measure of reproducibility of the measurement process and is the one that should be used to describe the precision of methodology and in the calculation of confidence limits when reporting analytical results.

The reader should be cautioned that the terms repeatability and reproducibility are used in a somewhat different sense than above on some occasions, and particularly where International Standards Organization guidelines are followed. The convention followed is:

$$\text{Repeatability} = 2\sqrt{2} \times \text{short term standard deviation}$$

$$\text{Reproducibility} = 2\sqrt{2} \times \text{long term standard deviation}$$

These represent the 95% confidence for agreement of two measured values over a short interval and a long interval of time, respectively.

Bias

When a measurement system is in statistical control and after its standard deviation has been evaluated, measurement bias can be evaluated most conveniently by the analysis of appropriate standard reference materials [134]. (See also Chapter 17.) When suitable reference materials are not available, comparison of the method under evaluation with one of known accuracy is another approach [135]. Any statistically significant difference between the known value of the reference material (or that of the comparison method) and a measurement can be attributed to bias. The reason for the bias should be sought. It may be inherent in the methodology or due to some artifact of the measurement process such as contamination, loss, or a calibration error. In the latter cases, corrective actions, including modification of the quality control program, should be taken. Only bias due to the methodology itself should be included as a figure of merit. It should be remembered that biased results can be obtained from unbiased methods of measurement, due to problems such as those mentioned.

Bias should be evaluated for each level of analyte measured in order to understand its nature and source. This is discussed in some detail in Chapter 24.

Selectivity

Selectivity refers to the uniqueness of response of a methodology. Rarely will a method respond only to a single analyte. Accordingly, it is important to know to what substances it may be responsive. The ideal situation is one in which potential respondents (interferants) are not present in the analytical sample or are present in such low concentrations as to cause no problems. When this is not the case, an interferant may be selectively removed, complexed, or masked by a chemical treatment. Improvement of resolution of the analytical system is another approach to solving problems of interference. Obviously, reliable measurements depend on the elimination of problems of interference and may require the use of alternate methods in extreme cases.

Selectivity is difficult to quantify and depends on the relative response to analytes of interest. The net effect is related to the response factors and to the relative concentrations of the analyte and the interferant. Depending on the accuracy requirements for the analytical results, the same interferant could cause a minor to major problem. In complex situations, experimental verification may be the only way to evaluate the magnitude of interference problems.

The error, E_n, for measurement of the nth component, due to interference, may be estimated using the relation:

$$E_n\ (\%) = 100\ \left(\frac{\Sigma R_i C_i}{R_n C_n} - 1\right)$$

where R_i represents the response factor for each analyte of concern under the conditions of measurement (all expressed in the same units), and C_i represents their respective concentrations in the analytical sample.

Other Considerations

Once it has been ascertained that the essential characteristics of the methodology are adequate for or exceed the data quality requirements, the characteristics listed as "desirable" in Table 9.2 may be influential in the selection process. Cost and ease of operation, including the competence level required for the staff, are very important. Some methodologies and especially instrumentation are more useful for research than for routine use. For example, versatility of measurement may be achieved at the expense of cost, ruggedness, and portability. When in doubt, the advice of similar users of proposed methodology or equipment should be sought on matters concerned with its practical use.

IMPORTANCE OF AN SOP

A well-developed procedure is necessary in order to use a method effectively. Furthermore, this must be optimized and used consistently if statistical measurement control is to be realized. Accordingly, an SOP should be used whenever measurement data are to be obtained. "Standard" means that the way the operation is to be done on each occasion is specified, which may or may not be a procedure developed by a standards-writing organization. However, when such are available, laboratories are well advised to consider them since they represent peer judgement and provide a basis for comparability of data among user laboratories. In the absence of an SOP, the analyst may employ "best judgment" and vary certain steps or operations, consciously or unknowingly, during a series of measurements. This could produce increased variability and even raise questions of the compatibility of individual measurements in a data set. When methods are used repeatedly, there may be sufficient incentive to write them up in a formal manner, but those used on an infrequent basis are often based on an analyst's "know-how." Such methodology is seldom optimized and frequently is used differently on each occasion of use. Accordingly, the use of unwritten methods is not considered to be good practice.

A well-written SOP will contain explicit instructions for carrying out the method (the procedure) and the tolerances for all critical measurement parameters that must be maintained in order to obtain results of a specified accuracy.

RUGGEDNESS TESTING

The critical operational parameters and the tolerances for their control need to be known for every measurement procedure if it is to be used intelligently. Such parameters include the environmental factors of temperature, pressure, and relative humidity, and chemical factors such as concentration of reagents, pH control, and the voltage and frequency of electrical inputs. Functional relationships for these should be known, or their relative effects on measured values should be estimated. With this information, one can estimate tolerances within which the parameters must be maintained in order to obtain results within acceptable limits. While possible to optimize such variables, this may not be feasible in many cases so that empirical operational values and their tolerances may be all that is possible. When developing methodology, it is

Table 9.3. Eight Combinations of Seven Factors Used to Test the Ruggedness of a Measurement Procedure

	Combination or Determination Number							
Factor Value	**1**	**2**	**3**	**4**	**5**	**6**	**7**	**8**
A or a	A	A	A	A	a	a	a	a
B or b	B	B	b	b	B	B	b	b
C or c	C	c	C	c	C	c	C	c
D or d	D	D	d	d	d	d	D	D
E or e	E	e	E	e	e	E	e	E
F or f	F	f	f	F	F	f	f	F
G or g	G	g	g	G	g	G	G	g
Observed result	s	t	u	v	w	x	y	z

Procedure

1. Choose 4 minus 4 combinations of s to z to get 4 caps minus 4 l.c. of desired letter.

example $$\frac{s+t+u+v}{4} - \frac{w+x+y+z}{4} = A - a$$

2. Rank the seven differences to identify problems
3. Calculate s from the eight results, conventionally, or

$$s^2 = 2/7\ \Sigma d^2$$

where $d = A - a$, e.g.

important to know whether the tolerances established are effective and/or whether unsuspected interactions exist.

Youden [158] recommended the use of a novel experimental plan, based on n + 1 measurements to test the effect of n variables, to determine whether established tolerances are effective (e.g., if $\pm$ t represents the tolerance for temperature control). In essence, he suggested that measurements be conducted in which the various critical parameters are maintained at either their upper limit or their lower tolerance limit in combinations as illustrated by the matrix of Table 9.3, designed for the study of 7 variables (factors) using 8 measurements. Thus measurement No. 1 is made with all variables at their maximum values (capital letter). Measurement No. 2 has variables A, B, and D at their maximum values while the remainder are at their minimum values.

Ruggedness testing is recommended when developing every SOP which has not been tested in this way. Furthermore, it should precede every collaborative test of a proposed standard method.

STANDARDIZATION OF METHODS

A number of organizations develop standard methods of chemical analysis, some in narrow fields and others in broad areas of measurement. The result is usually a procedure written in a standard format; hence, a standard method is essentially an SOP.

The standardization processes used by various organizations differ, and this should be borne in mind when considering the use of a standard method. Also, the formats used for the publication of the standards differ, and one may be more convenient than another in some cases. Figure 9.4 illustrates several pathways that can be followed in standardization of methodology. The simplest process consists in the collection and selection of methods, based on the judgment of a technical expert(s), after which they are written in a standard format and subjected to editorial review by a committee, for example. The most extensive process involves collaborative testing which may or may not be preceded by a peer laboratory test. In this case, the result is a fully evaluated method, sometimes called a validated method. Readers are referred to the papers, "Validation of Analytical Methods" [135] and "The Role of Collaborative and Cooperative Testing" [142] for further discussion of the significance of such testing.

The procedures used in collaborative testing vary from organization to organization but follow the basic plan shown in Figure 9.5. When a peer or coordinating laboratory is not used, the functions ascribed to it may be done by a responsible committee, or some of them may be assigned to one or more of the participants as appropriate.

The process followed and the statistical treatment of the data in the case of AOAC [159] and ASTM [9, 18] are described in the references. Essentially, a group of laboratories analyze appropriate test samples. If the methodology encompasses a wide range of concentrations, test samples representing the low, middle, and upper range of concentrations are recommended as a minimum.

The test samples should be chosen for their typicality to those for which the methodology is intended, and their stability and homogeneity are major considerations. The compositions (and homogeneity) should be known before use, otherwise only consensus values of the test results can be obtained which would provide only limited if any evaluation of bias. When the methodology has been thoroughly researched, as is highly recommended, bias of the methodology does not need to be evaluated in the collaborative test, and this puts less constraints on the test materials. Ruggedness testing by a peer laboratory should precede any collaborative test.

The participating laboratories should engage in sufficient familiarization measurements, prior to the collaborative test, to develop sufficient expertise in the methodology and demonstrate statistical control. The collaborative test should not involve a learning experience. The objective of the test should be to

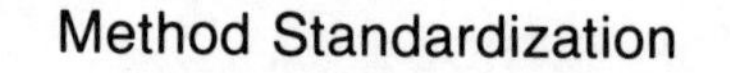

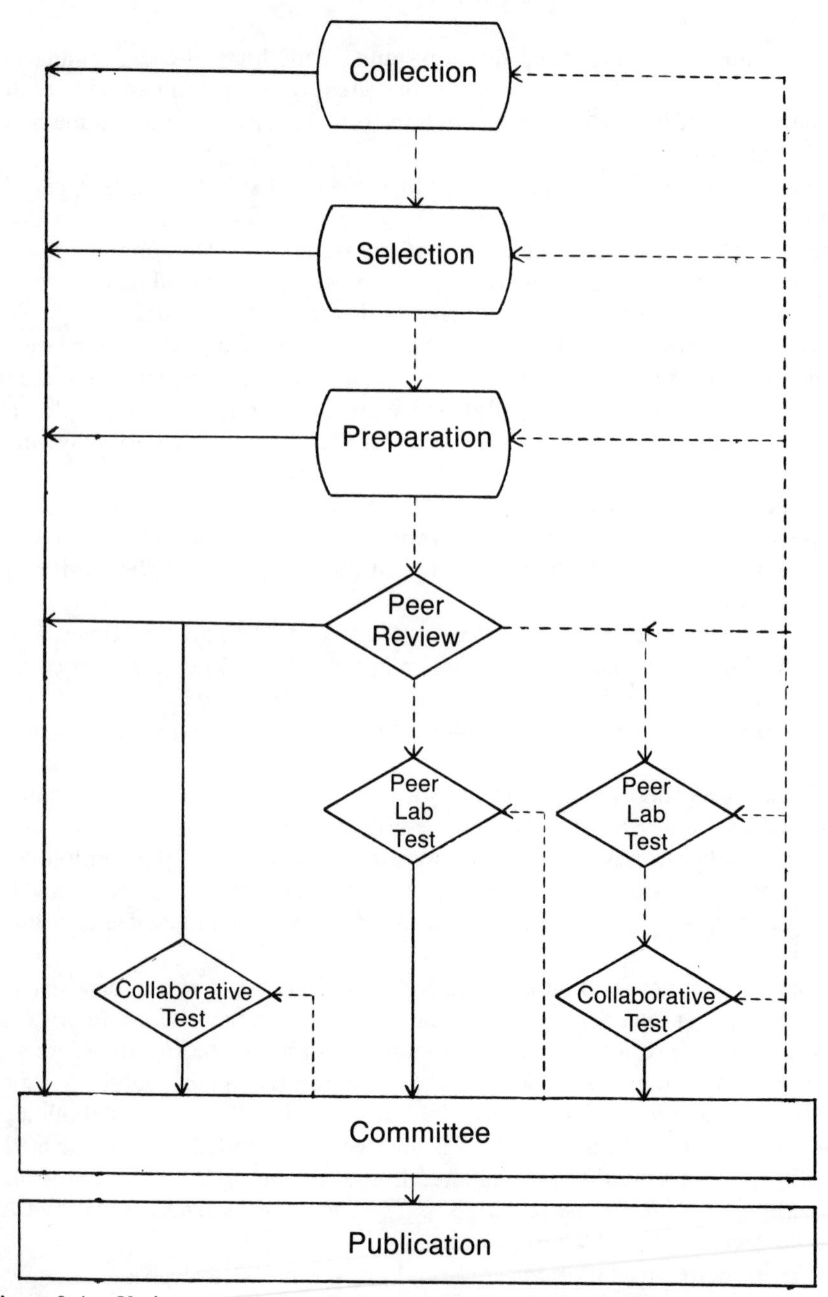

Figure 9.4. Various approaches used to standardize methods of measurement.

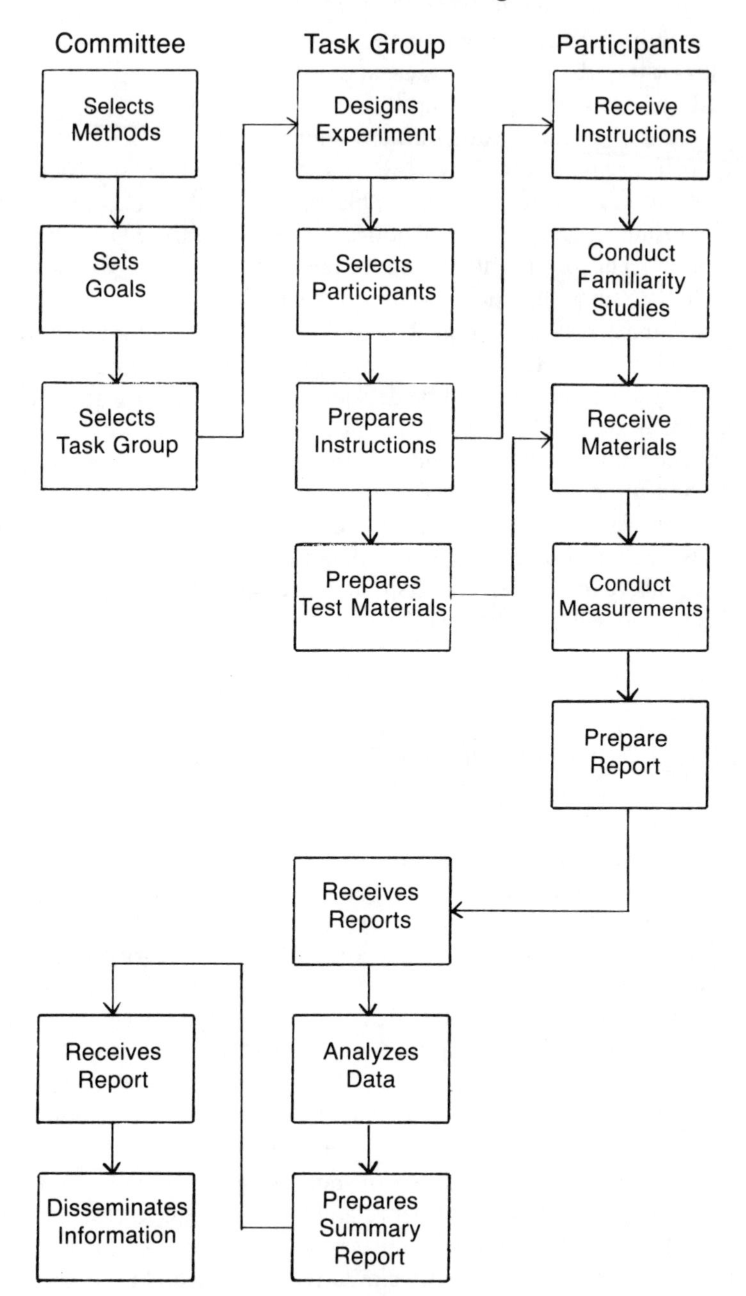

Figure 9.5. Procedures used for the collaborative testing of methods of measurement.

confirm usefulness of the methodology in the hands of competent analysts who are thoroughly familiar with the test method.

The collaborative test consists of the multiple analysis of samples at several levels. From this data, the between and within laboratory precision can be evaluated, as well as significant biases between laboratories. Laboratory results should be examined for outliers—not to improve the statistics, but to identify measurement problems. All outliers as well as significant biases should be investigated to find assignable causes for their occurrence. Where this is the result of defects or ambiguities in the procedure (SOP), these should be corrected. The organization/committee developing the standard should decide whether such an action requires a further testing of the method.

Each standard method should contain full information on the figures of merit related to it, and these are best developed by a peer laboratory. Precision and accuracy statements should be developed and prepared with care. Unfortunately, such statements are often misinterpreted. They only describe the results of the collaborative test and may be typical, only if one can believe that the participating laboratories were a random sample of possible users of the methodology. Every laboratory must evaluate its own statistics for use of a method, which may be better or worse than those obtained during the collaborative test. These and other matters concerned with the interpretation of test results are discussed elsewhere [142].

SELECTION OF APPROPRIATE METHODOLOGY

Figure 9.6 shows the essential steps that should be followed when selecting methodology for analytical measurements. Based on an understanding of the problem, the data requirements can be established. The methodology available will fall into one of the several categories shown. Category A is the most useful, and its use requires verification only of the quality assurance aspects and especially attainment of statistical control of the measurement process before measurement of test samples. Category B differs only slightly from A in that the additional steps of equipment check-out and the development of an SOP are required before proceeding to analyze samples. When completed, this will move this methodology closer to, if not into, category A.

A study of the figure will indicate the additional research and development that may be required as the confidence and/or knowledge of the feasibility of any method under consideration decreases. It will be noticed that two steps are common to all categories. An SOP should be available and used, and all other aspects of quality assurance should be observed. The SOP will facilitate attainment of the statistical control of the present measurement process and facilitate future use of the methodology.

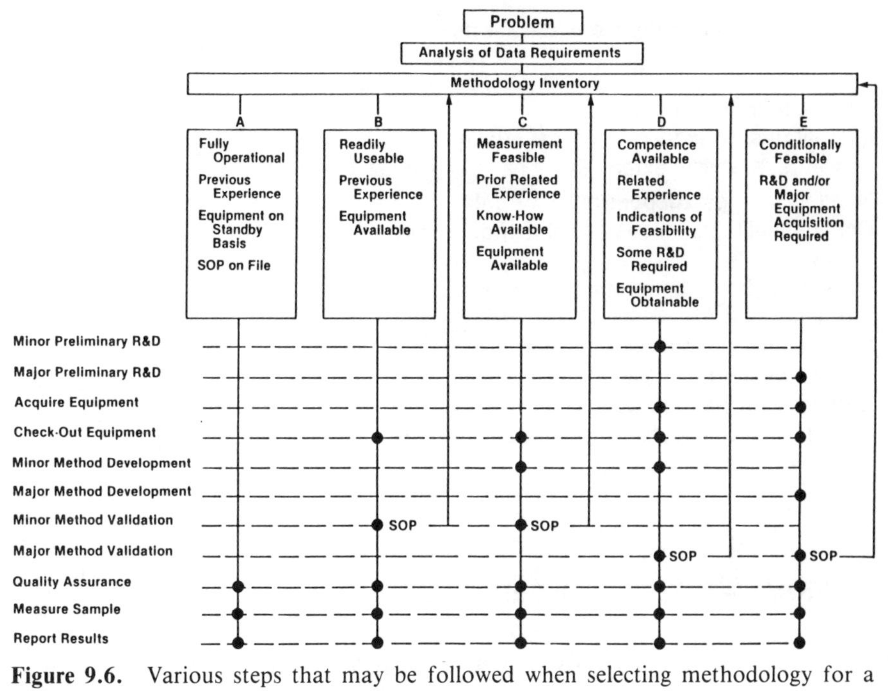

Figure 9.6. Various steps that may be followed when selecting methodology for a specific use.

GUIDELINES FOR SELECTING INSTRUMENTATION

Most modern analytical methods involve instruments for some or all of the steps. The selection of reliable instruments is definitely a part of quality assurance. The following items should be considered when choosing specific components or complete measurement systems.

- Validity of methodology
- Versatility
- Ruggedness and simplicity of design
- Skill requirements
- Compatibility with existing equipment
- Availability of service and parts
- Reputation and experience of manufacturer
- Capability for updating or adding accessories in the event of changes of technology or in the program of the laboratory

Table 9.4. Feasibility Considerations

	Yes	No
Favorable in-house cost	___	___
Internal quality exceeds external quality	___	___
Limited availability, externally	___	___
Need for internal responsibility for data	___	___
Confidentiality questions possible if done externally	___	___
Turn-around time important and more favorable for internal work	___	___
Fills vital gap in internal program	___	___
Service needed to attract other services	___	___
Loss-leader considerations	___	___
Provides full employment with resulting losses compensated by other gains	___	___

INTERNAL VS EXTERNAL SERVICES

Modern chemical analysis is complex and often requires special expensive facilities and equipment, sometimes with high obsolescence potential. The question of whether to tool up for in-house services or to rely on a specialist laboratory is one of increasing concern. For such decisions, a realistic cost estimate is the first requirement. An expression that can be used to estimate the unit cost of analysis is as follows:

$$\text{cost/analysis} = \frac{0.33\ \text{IC} + \text{MC}}{\text{D}} + \text{DL} + \text{OH}$$

where IC = initial cost of equipment
MC = annual maintenance cost
DL = direct labor cost/per analysis
OH = overhead cost chargable/per analysis
D = expected demand for the service, number of analyses/per year

The factor 0.33 is based on an amortization of equipment cost over a 5-year period at an interest rate of 11%. The maintenance costs include those for supplies, parts, servicing, and special facilities not included in overhead. In making the direct labor estimate, the cost of calibration, sample preparation, calculations, report writing, and cleanup should be included. If these operations are done by different individuals with various hourly rates, this needs to be considered.

Overhead costs, OH, consist of all indirect costs not specifically chargeable to a particular analysis, prorated on an annual basis. This is a maximum cost.

A minimum cost would be based on substituting full capacity utilization, CU, for D. The value for CU is calculated by:

$$CU = \frac{2000 \times N - \text{down time}}{\text{instrument hours/analysis}}$$

where N = number of shifts per day

It is recommended that a 10% markup be made and utilized for quality assurance (reference material analysis, control charts, etc.) unless already included in analysis time. Profit and/or contingency factors should be applied as appropriate. No costs of training of personnel are included. There is a minimum demand that will make a given measurement technique feasible.

Even with a cost estimate in hand, the final decision still could be subjective. A suggested list of items of this nature for consideration is given in Table 9.4. A favorable cost estimate would seem to be decisive, but this should be based on reasonable assurance of a continuing need. In some cases, it may be necessary or desirable to establish an in-house capability even in the event of unfavorable costs as suggested by the other items in the checklist. These items will need to be weighted depending on their perceived relative importance. Obviously, if one needs to proceed far down the list to encounter a "yes," such an item would need a high weight to merit feasibility. If no "yes" boxes are checked, the proposed methodology would need to be justified on the basis of prestige or other intangible factors which could be decisive in some cases.

Whenever a laboratory uses the services of another laboratory, it must ascertain that the vendor is providing the quality of data that is needed. Analytical services must be purchased under quality specifications just as any other commodity. Because of the difficulty of complete specification, and also because of the problems of monitoring for compliance, the quality assurance practices of a vendor are of prime importance. This subject is discussed further in Chapter 26.

CHAPTER 10

Principles of Calibration

The concept of calibration is very broad. As a noun, the word means a set of graduations marked to indicate values such as the markings on a thermometer. As a verb, it means the comparison of a measurement standard or instrument with another standard or instrument to report, or eliminate by adjustment, any variation (deviation) in the accuracy of the item being compared.

In chemical measurements, it refers to the process by which the response of a measurement system is related to the concentration or amount of analyte of interest, and hence often consists of the evaluation of an analog response function.

Standardization is a term related to calibration. As used by chemists, it refers to the establishment of the value of a potential chemical standard (e.g., a titrant), usually by comparison with a standard of known composition.

The concept of tolerance is also related to calibration. A tolerance is a deviation from the nominal value of a standard which is considered to be negligible with respect to its intended use. Thus, class tolerances are established, often on the basis of the consensus of users, and standards are considered to be within tolerance when their errors do not exceed them. Tolerance testing consists of checking for compliance, but no correction factors are assigned. Because the nominal value of the standard will be used, the user must ascertain that the tolerance is realistic for a specific application, since the standard could deviate from its nominal value by the total amount of the tolerance.

An analogous situation is the use of the nominal purity of a purchased chemical for preparation of chemical standards. The uncertainty associated with this is discussed later.

Regardless of the exact definition, the concept of calibration is basic to all measurement. Measurement is essentially a comparison process in which an unknown whose value is to be determined is compared with a known standard. The comparison may be direct, as in weighing with an equal-arm balance, or indirect using a previously calibrated instrument or scale. Even in the former case, the items compared are seldom identical so that some kind of calibrated instrument or scale is involved in evaluating the magnitude of any observed difference. Obviously, its calibration must be accurate if accurate measurement data are to be achieved.

WHAT NEEDS TO BE CALIBRATED

The only purpose of calibration is to eliminate or minimize bias in a measurement process. The precisions of calibrated and uncalibrated systems can be equivalent. Chemical measurements are made using a system that includes sampling and measurement processes. All aspects of the measurement process need to be calibrated. Likewise, all aspects of the sampling process that involve measurement in any way (sieving, for example, and even the locating of sample sites) should be calibrated. Ancillary data on such matters as temperature, pressure, humidity, particle size, volumetric capacity, mass, and flow rate may be needed as well, requiring accurately calibrated instrumentation for their measurement. Accordingly, any of the instruments, standards, and methods used for these purposes must be calibrated to assure that their accuracy is within acceptable limits.

REQUIREMENTS FOR CALIBRATION

The prime requirement for calibration is the availability and use of appropriate and accurate standards. Minimally, they should have a high degree of similarity, and ideally, they should be identical to the object compared. In physical measurement, this is readily achievable, but it is often difficult and sometimes approaches the impossible in chemical measurements. Even the effects of small deviations from matrix match and analyte concentration level may need to be considered and evaluated on the basis of theoretical and/or experimental evidence. Otherwise, accurate measurement data may be unachievable. Equally important is the accuracy of the standards used for calibration purposes. Chemical standards ordinarily are prepared by quantitatively combining constituents of known purity, but the latter cannot be automatically assumed. The purity of the constituent representing the analyte of interest must be known to at least the accuracy requirement for its measurement. When trace levels of an analyte are of concern, even minute amounts in a diluent matrix can cause large errors in a calibration standard unless accurate measurements are made and appropriate corrections are made for the analyte's presence. An analyst preparing his own standards should have the capability of evaluating them. If not, it is imperative to use only starting materials or standards prepared by suppliers of the highest reliability. In such cases, analysts should require proof of such reliability based on the supplier's quality assurance program.

Similarly, the stability of standards, however and by whomever prepared, is a prime requirement. Every standard should have an assigned expiration date indicating its stable life expectancy and should not be used beyond such a date.

FREQUENCY OF CALIBRATION

An important aspect of calibration is the determination of the calibration interval, i.e., the maximum period between successive recalibrations [65]. Two basic and opposing considerations are involved: the risk of being out of tolerance at any time of use, and the cost in time and effort of calibration. The former should be the major concern because of the dilemma of what to do with data obtained during the interval between last known in and first known out of calibration. However, an overly conservative approach could be prohibitively expensive. A realistic calibration schedule should reduce the risk of the former without undue cost and disruption to work schedules. The factors that need to be considered in a realistic calibration schedule include:

- accuracy requirement for the measurement data
- level of risk involved
- experience of the laboratory in use of the equipment or methodology
- experience of the measurement community
- manufacturer recommendations
- external requirements for acceptability of data
- cost of calibration

An initial choice of calibration interval may be made on the basis of previous knowledge or intuition. Based on experience gained during its use, the interval could be extended if the methodology is always within tolerance at each recalibration, or it should be decreased if significant out-of-tolerance is observed. Control charts may be used to monitor the change of the measured value of a stable test item correlated with the need to recalibrate. If a sufficiently large number of test instruments is involved, as in a monitoring program, a statistical study may be used to establish calibration intervals for critical items or procedures.

MODE OF CALIBRATION

Calibration of standards (perhaps standardization is a better term) is a static process. That is to say, their values should be independent of who performs the calibration. Conversely, calibration of measurement systems is a dynamic process in that both the method of use and the user are involved. First, the system must be in statistical control since calibration's only function is to minimize bias. Then the question arises as to how much of the system to calibrate—the sensor, the measurement process, or the entire analytical system. The answer will depend on the sources of bias that need to be calibrated and the calibration interval required for each. Ordinarily, the response of the sensor and the related instrumental analogue function will need frequent calibration which is the easiest operation to perform. Often, only the introduction of a known gas mixture, test specimen, test solution, or a series of such is all that is required for this purpose. If judgment of the analyst is involved in a measurement process, a separate calibration could be required for the operation of the system by each analyst that will use it.

The entire analytical system should be calibrated for each class of test samples, perhaps less frequently than the individual components, but never ignored. Calibration materials consisting of reference materials closely resembling the test samples will be needed for this purpose but may be difficult or impossible to prepare or obtain. The analyst may be forced to use surrogates added to the test sample or the method of standard additions of analyte to calibrate the measurement system. Because an artificially added analyte may not necessarily respond in the same manner as a naturally occurring analyte, the above approaches may need to be validated.

The analyst must bear the responsibility for the adequacy of all calibrations related to his measurements. Methods of calibration, especially for physical test equipment, are the subject of a number of ASTM standards, and these should be used whenever applicable. Each chemical method should contain explicit procedures for its calibration, but these are often limited to the measurement step or even to the sensor response. The calibration of the entire analytical system will require the use of appropriate standards as described. The analyst must bear the full responsibility for their appropriateness.

LINEAR RELATIONSHIPS

Many chemical measurements do not involve direct comparisons with standards but rather indirect and intermittent comparisons in which the only use of the standards is to establish an analytical response function that is used in subsequent analytical measurements. A linear function is typically appropriate, or instrument output may be linearized to achieve this condition. The results of measurement with an instrument having theoretical linear response are illustrated in Figure 10.1. When in a state of statistical control, repetitive responses to each standard are randomly distributed around limiting means (population means) that fit a straight line perfectly. In practice, a limited number of measurements are made to establish a response function so that limiting means are not realized but rather sample means which deviate somewhat from a straight line as shown in Figure 10.2. In such cases, the uncertainty of fit of the calibration function is a matter of concern [64].

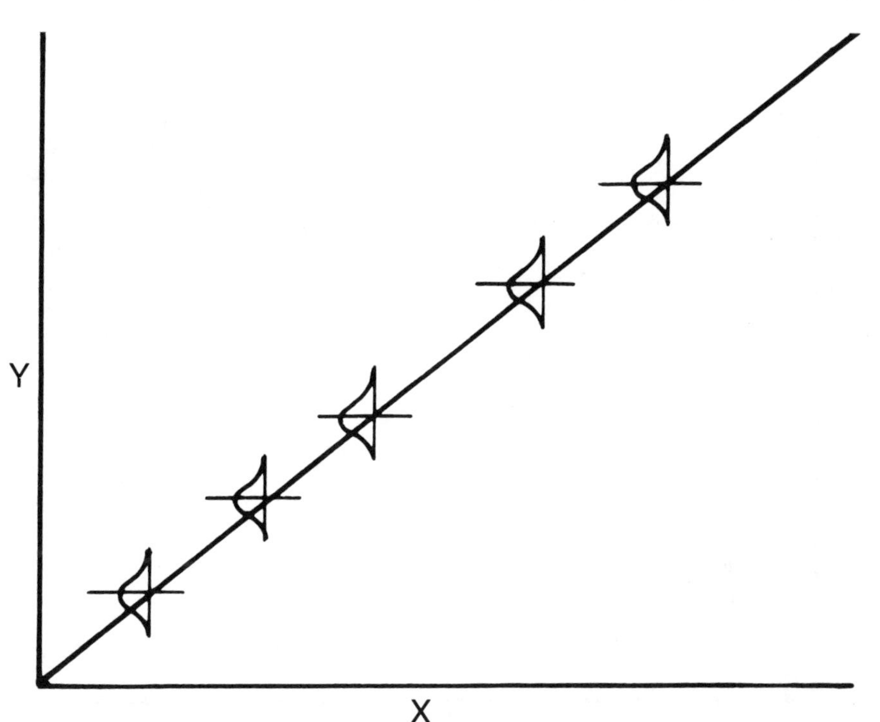

Figure 10.1. Linear fit of a theoretical relationship. The plot shows a perfect fit of the population of repetitive measurements of errorless standards.

In a typical chemical calibration situation, a series of calibration standards, considered to have insignificant compositional uncertainty, are measured one or more times. The response is then plotted, and a straight line is graphically drawn or fitted by the method of least-squares. The latter procedure is preferable in that it provides an unequivocal estimate of the uncertainty of fit. Also, the standard deviation of the slope and intercept can be calculated as well as the confidence interval for the line as a whole, a point on the line, and a single Y corresponding to a new value of X [100].

The joint confidence interval for the slope and the intercept can also be calculated to assist in deciding whether calibration lines obtained on various occasions are consistent (Figure 10.3) [64]. Because the slope and the intercept are correlated, only a small range of values for each are mutually compatible at a selected level of confidence. The values for 95% confidence do not fall within plus or minus two sigmas of the slopes and intercepts but within an ellipse as shown.

It has been recommended that any mathematical fit of data should be preceded by a graphical plot [100]. This will indicate whether a linear relationship is justified, and also it may be used to screen for outlying data points. Linear

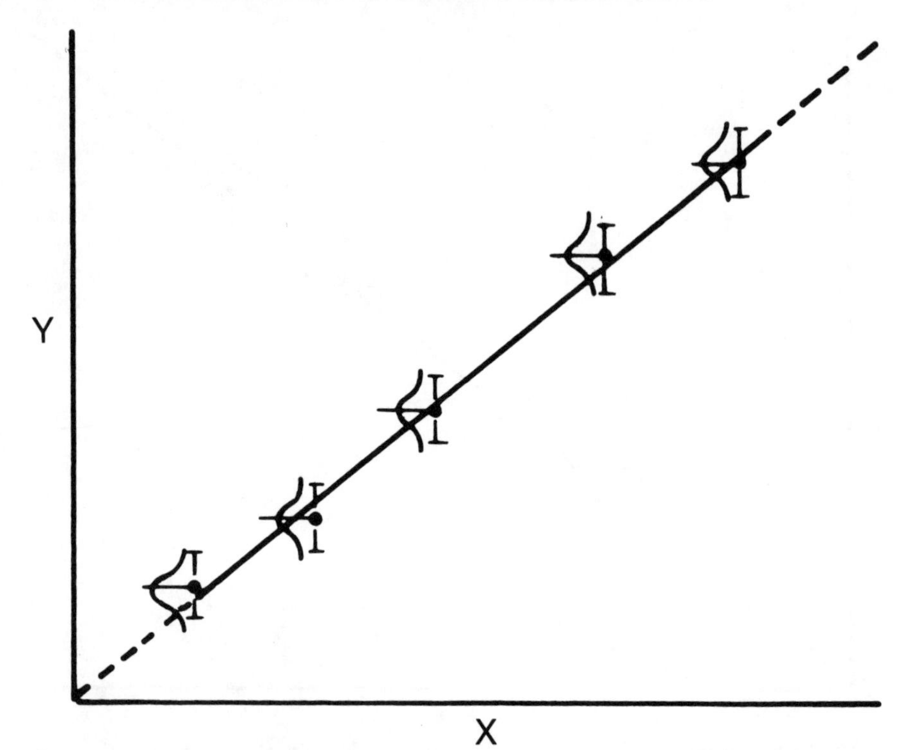

Figure 10.2. Linear fit of a limited data set. The plot shows the fit of a sample of measurements (even one) of errorless standards. There is an uncertainty of fit due to measurement uncertainty, and possible to questions of linearity.

regression will try to accommodate all points by minimizing the sum of the squares of the residuals. The presence of outliers, ignored in a graphical plot, will be influential in a least-squares fit and can compromise an otherwise excellent data set.

The uncertainty of the linear fit should be added in quadrature to that of the measurement process when overall uncertainty is being estimated. If calibration uncertainty is statistically significant and undesirably large, it can be reduced by multipoint and/or multimeasurement calibration.

In some calibration situations, such as ultratrace analysis, the uncertainties of the standards used may not be insignificant as illustrated in Figure 10.4. In such cases, the simple regression fit (x-axis essentially errorless) is not justified [87]. The assistance of mathematical or statistical experts knowledgeable in handling such data may be required.

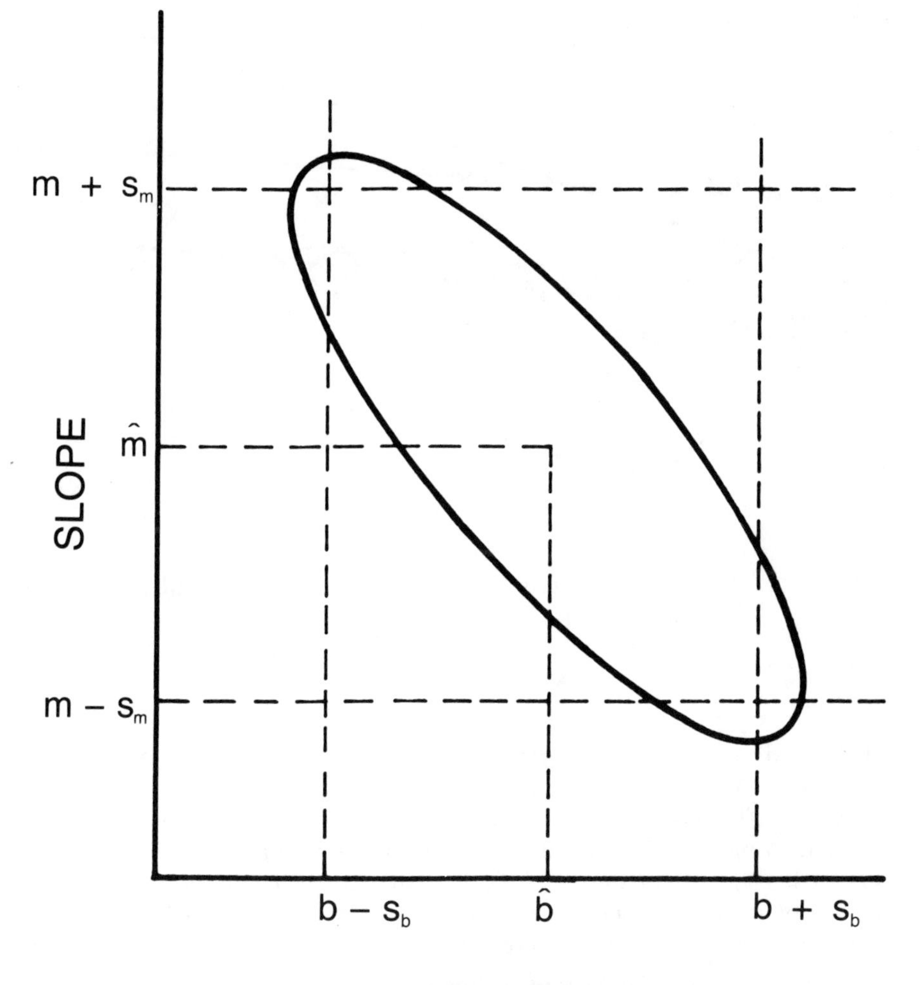

Figure 10.3. Joint confidence ellipse for slope and intercept. Because the slope and intercept are correlated, there are only a limited set of values, (those contained within the ellipse) that are mutually compatible.

TESTS FOR LINEARITY

Either direct or transformed data may be represented by a functional relationship of which the linear model is the simplest. There are several approaches that may be followed to decide whether such a model is valid.

If there is a theoretical basis for a linear model, it should fit the data. If it does not, there could be a bias(s) that causes deviation, and this should be investigated and appropriate corrective action(s) taken.

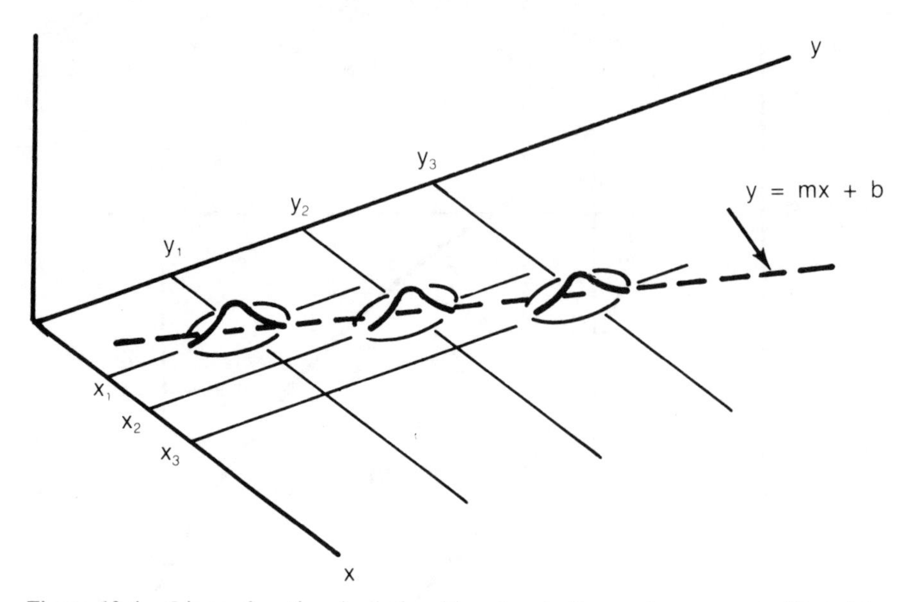

Figure 10.4. Linear functional relationship when both x and y axes are affected by measurement errors.

Empirical apparent linear relationships can be confirmed by several approaches. A plot or fit of the data may appear to be linear in that no systematic differences between the data and its functional (or graphical) fit are observed. A plot of the residuals (Δ = observed − computed) with respect to the variable should not show significant systematic behavior. If a reasonable number of data points (at least 7) are available and more or less equally distributed over the range of interest, this can be a useful approach.

When a small number of data points are available, the above test is less conclusive, in which case a probability of fit may be required. Systematic trends that may be present may not be significant, and apparently good fits can occur when the linear relationship is not valid.

A simple statistical test (the F test; see Chapter 4) may be used to provide guidance in the situation above. The standard deviation of measurement at each point (s_w), based on multiple measurements of standards, is needed. The values for s_w at the several calibration points may be pooled to obtain an estimate based on an increased number of degrees of freedom. The value of the variance, s^2_w, is compared with the variance, s^2_b, of the points about the fitted function to obtain a value.

$$F = \frac{s^2_b}{s^2_w}$$

In this case, ν for s^2_b is equal to n - 2, where n is the number of fitted points, while ν for s^2_w represents the degrees of freedom of the pooled standard deviation estimate, s_w. If the value of F is significant, s^2_b may be considered larger than s^2_w which suggests that the fit is not consistent with the scatter of the points. Unfortunately, judgment of departures from linearity based on a small number of data points are difficult to make due to the insensitivity of the F test.

A graphical variation of the above approach may be used based on drawing error bars (e.g., 95% confidence bars) for each plotted point. If these bars intersect the fitted line in a random manner, a linear fit probably is justified.

Nonlinear data can approximate a linear fit if the range is suitably small. Accordingly, in some cases, it may be advantageous to group the data in several subranges, each of which is practically linear.

For the most accurate measurements, the point measured may be bracketed by two closely spaced calibration standards. Such a calibration becomes an interpolation between the standards, hence considerations of linearity virtually vanish.

Correlation coefficients are sometimes recommended for judgement of fit of an equation to a data set. While correlation coefficients are good tests for correlation, they are not necessarily good tests for linearity. The tests discussed above are considered more useful for this purpose.

CALIBRATION UNCERTAINTIES

Ideally, the calibration process is undertaken to eliminate deviations in the accuracy of measurements or instruments. However, this cannot be glibly assumed. In fact, as the limits of measurement are approached, the uncertainties of calibration may increase in a similar manner and can be the limiting factor in attainable accuracy.

Uncertainty of calibration may be characterized according to the confidence in the standards used and in the uncertainties of their use in the measurement process. The uncertainty in the composition of chemical standards will depend on the degree of experimental realization of the calculated composition based on knowledge of the purity of constituents, on the accuracy of the preparative process, and on considerations of stability. The reliability of the process for transferring the standard to the system calibrated is a further consideration. Both systematic and random sources of error are involved in all of the above and will need to be minimized to meet the accuracy requirements of the data. Repetitive calibrations will decrease the random component of uncertainty but not any biases. As calibration uncertainty and measurement uncertainty approach each other, calibration can become a major activity, even in routine measurements.

The statistical uncertainty of fit of a calibration line may be shown

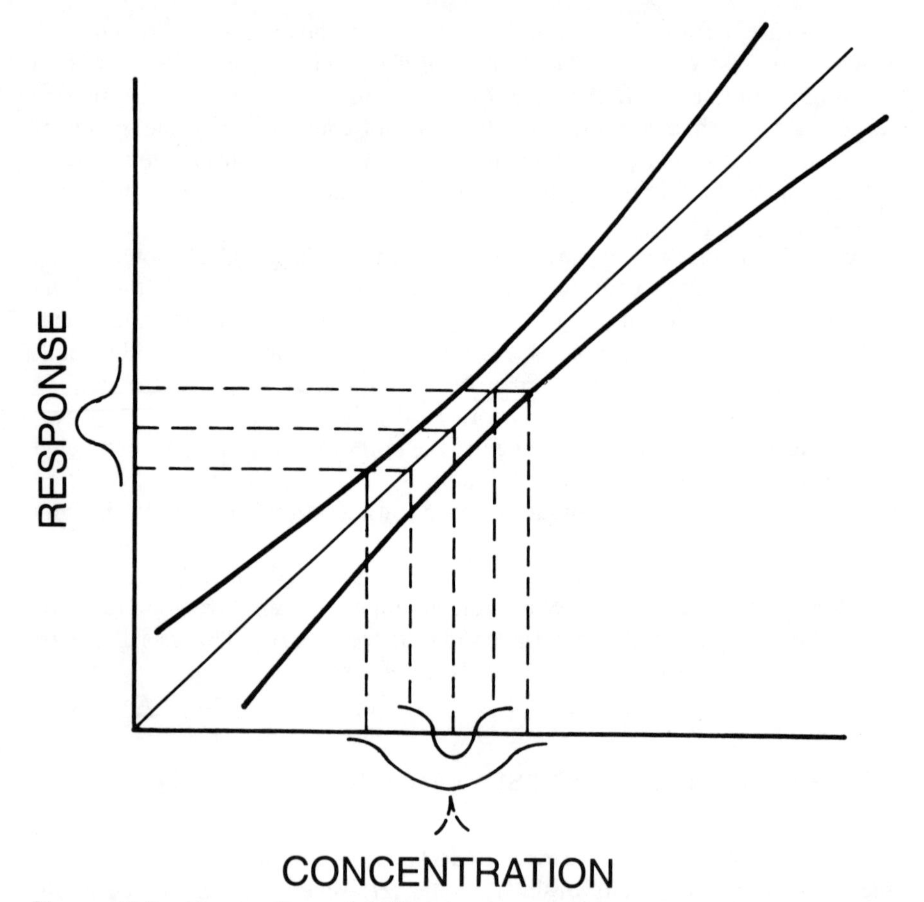

Figure 10.5. Propagation of calibration uncertainty. The uncertainty of the calibration line is shown as a band of varying width. This uncertainty adds (in quadrature) to measurement uncertainty when considering the total analytical uncertainty.

(Figure 10.5) as a band surrounding the fitted line. The procedure for constructing such a band is given by Natrella [100]. In the case of equispaced data, the band is narrowest at the center of the plot and broadens at the extremities. The increased confidence for the center arises because it has the highest number of degrees of freedom, which decrease in each direction from the center. The lower confidence when using such a line in an extrapolation mode is a consequence of this as well as the concern for linearity of the calibration data over an extended range. The figure shows qualitatively how the uncertainty of the calibration line adds to measurement uncertainty each time it is used.

The uncertainty of a calibration line may be decreased in several ways. Increasing the number of calibration points narrows the band, as does increasing the number of *independent* measurements of each calibration standard.

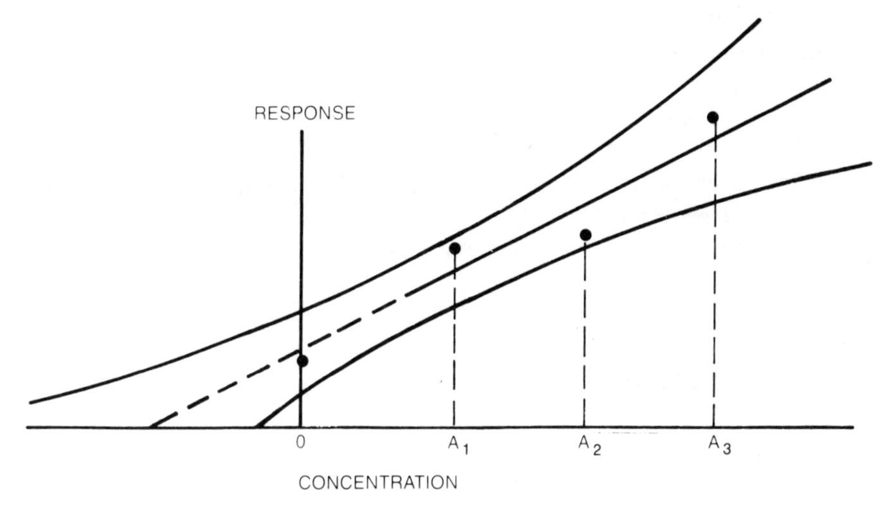

Figure 10.6. Uncertainty of the spiking mode of calibration.

ing the number of *independent* measurements of each calibration standard. The practice of using the line in its region of smallest uncertainty, whenever possible, is another way to decrease calibration uncertainty.

The uncertainty of calibration using the method of standard additions (spiking) is illustrated in Figure 10.6. If only two spikes are added, such as A_1 and A_2, no estimation can be made of the uncertainty of the resulting calibration line. Even if three spikes are added, the width of the uncertainty band is large since only one degree of freedom is involved in fitting the line. Moreover, the uncertainty of C_o is large because it is obtained by extrapolation.

The dilemma in choosing spiking levels is also illustrated in Figure 10.6. If only two levels are chosen, and they are close together, small experimental errors in measurement of A_1 and A_2 can make relatively large errors in $\Delta y = y_{A_2} - y_{A_1}$ and hence for the slope $\Delta y/(A_2 - A_1)$. On the other hand, one always prefers that the spikes should be close in level to that of the analyte in the samples measured. The spacing shown in the figure is reasonable for most situations.

In addition to the statistical uncertainty just described, spiking may have technical uncertainties due to whether an artificially added analyte can be recovered with the same efficiency as a natural one. Accordingly, spiking should not be used for calibration and validation when other approaches are feasible.

To all of the above must be added any uncertainty related to the appropriateness of the standards. If accurate standards could be prepared, identical or with insignificant dissimilarity to the unknowns, measurement would be simplified in that the analytical uncertainty would depend entirely on the precision of comparison. Only in rare cases is this possible. The complexity of most matrices (natural or manufactured) and the uncertainty (and virtual impossibility) of duplicating the physical and chemical interrelations of the components of the matrix impedes the synthesis of equivalent standards. The sheer number of standards that would be required makes such an approach impractical even if it were possible to achieve. Accordingly, chemical calibration is an approximation at best. The analytical chemist must be constantly aware of the possibility of bias introduced by the nature of the standards used, which may be the major source of bias in the analytical data. Appropriate reference materials should be used to evaluate this and other aspects of the measurement process [126].

INTERCALIBRATION

Intercalibration may be defined as the process, procedures, and activities used to ensure that the several laboratories that contribute data for a common purpose can produce compatible data [128, 138]. The concept is applicable to individuals within the same laboratory, satellite laboratories within the same organization, and participants in a monitoring program. Such laboratories should implement a reliable and consistent calibration program and achieve statistical control before any intercalibration program can be meaningful. Intercalibration is best evaluated by systematic use of standard reference materials (SRMs) or similar substances. The reference materials should have a high degree of comparability with the test samples of the monitoring program, since the confidence in the measured values of the latter will be inferred from that of the former.

While at least one test level is necessary, this may not be sufficient in that it can only demonstrate intercalibration at that level. A minimum of three test levels (five is desirable) spanning the measurement range is necessary to evaluate the performance and to intercalibrate a measurement system fully. (See Chapter 24 for a further discussion of this subject.)

ADDITIONAL REMARKS

By the very nature of measurement, calibration is one of its most critical steps. While the uncertainty of measurement can be worse than that of calibration, it can never be better. When statistical control has been verified, calibration uncertainties can be the remaining major source of discrepancies between laboratories. Thus, biased measurement data can result when using unbiased methodology because of calibration bias. Laboratories are urged to give more attention to this important operation and to critically evaluate it in every measurement situation. The consistent use of appropriate reference materials together with control charts can assure laboratories that unacceptable biases are not present in their measurements of calibration.

CHAPTER 11

Principles of Quality Assurance

Up to this point, the emphasis of this book has been on the basic principles of the chemical analytical system. The chapters that follow are concerned with the attainment and maintenance of adequate and acceptable performance of the system and with the evaluation of its outputs. The premise has been stated earlier that the quality of data must be known before it can be used in a logical sense and that data of consistently acceptable quality is dependent on appropriate quality assurance practices. When all is said and done, a quality assurance program is conducted:

- to discharge management's responsibility for the quality of a laboratory's outputs
- to allay the analysts' concerns for the quality of their work
- to inform the user of the quality of data or services so that they may be used with confidence
- to provide records and documentation for present and future use
- to protect all of the above interests

In the production of useful items, the judgment of quality is based primarily on how well some need has been met. This judgment involves:

- knowing the user's needs
- designing to meet these needs
- faultless implementation of production plans
- reliable performance of subcontractors

In addition to the above, quality may be judged by the intangible qualities of:

- punctual delivery
- prompt and efficient backup services

All of the above concepts of quality apply in principle to measurement services and should be considered in a generic sense when reviewing, revising, or developing the quality assurance procedures to be followed by a laboratory. In addition to the above, quality of data may be influenced by the quality of reporting it. This has some parallelism to the user's manuals often supplied with devices or equipment. Well-written manuals and reports will enhance the usefulness and discourage the disuse of both the item and data.

Modern quality assurance practices are based on the philosophy that defects can be prevented and that the major purpose of detection is prevention. A reliable laboratory must be dedicated to never ending improvement of quality and productivity [43]. The philosophy must be:

The goals of yesterday
are the commonplace occurrences of today,
and the outmoded practices of tomorrow.

In preview, quality assurance may be considered as composed of the related activities shown in Figure 11.1. Not only must the kind and amount of quality assurance activity be considered in relation to the aims and objectives of the

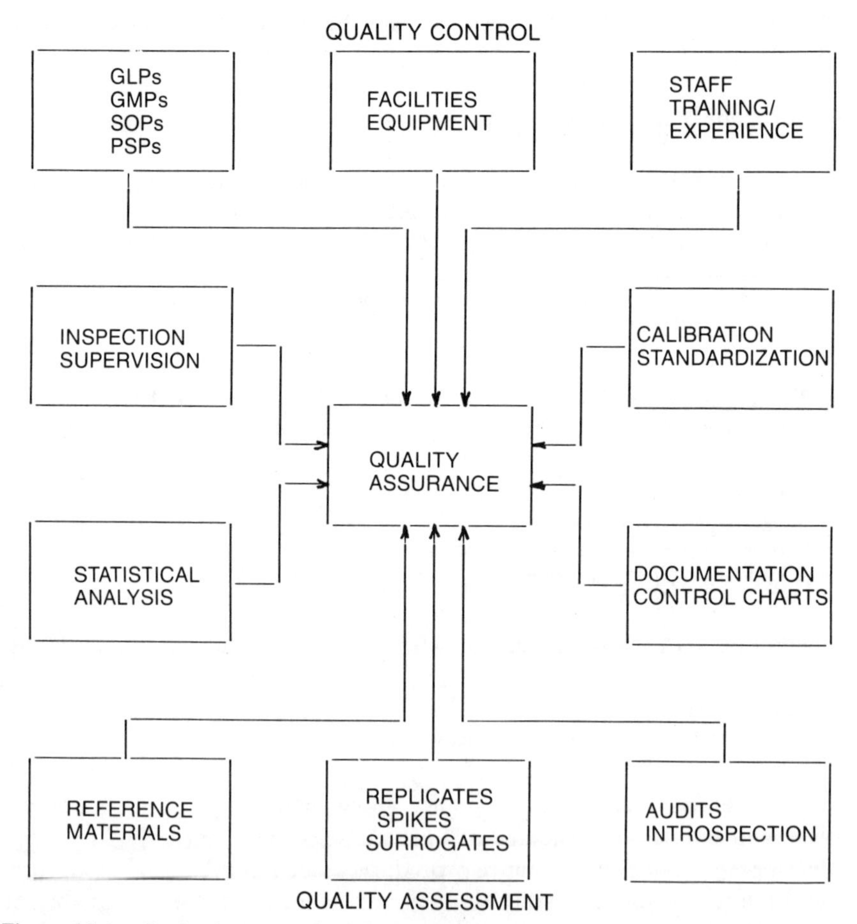

Figure 11.1. Basic elements of quality assurance.

measurement program, but a balance must be achieved between the amount of effort devoted to each of its two component parts.

The principles of quality control and quality assessment are discussed in some detail in the following chapters. Specific details of the operations involved are then discussed. With such knowledge, a laboratory should be able to make intelligent decisions as to what and how much quality assurance needs to be implemented in its general program of work and for specific applications as they occur.

CHAPTER 12

Principles of Quality Control

Quality control techniques include all practices and procedures that lead to statistical control and to the achievement of the accuracy requirements of the measurement process. The basic elements of quality control are listed in Table 12.1. The list proceeds from the general to the specific, and each element depends on the ones above it [137].

COMPETENCE

Adequate competence and expertise of the technical staff are the first requirements for quality measurements. Despite the increasing sophistication of modern instrumentation, technical judgment, experience, skill, and even the professional attitude of the staff are very important for reducing and maintaining measurement variability to acceptable levels. The laboratory staff must have at least a minimum level of competence commensurate with the skill requirements of the analytical discipline in which they are engaged. Only in the most repetitive measurement situations can the chemical measurement process be reduced to a mechanical routine. It is common knowledge that measurement competence increases as experience is gained with most methods and that laboratories, seemingly equal with respect to facilities and equipment, can differ significantly in the quality of their data outputs.

Table 12.1. Basic Elements of Quality Control

Technical competence of staff
Suitable facilities and equipment
Good laboratory practices (GLPs)
Good measurement practices (GMPs)
Standard operations procedures (SOPs)
Protocols for specific purposes (PSPs)
Inspection
Documentation
Training

Competence is based on an adequate and sound educational background, specific training to perform assigned duties, and experience in the use of the techniques and methodologies employed. Attitude of staff is a very important contributor.

SUITABLE FACILITIES AND EQUIPMENT

Modern chemical analysis typically requires specialized facilities, apparatus, and equipment, and success or failure can often be traced to their adequacy and mode of utilization. Thus, ultratrace analysis can hardly be done without ultraclean laboratories, and toxic substance analysis may require the use of containment areas. Humidity and temperature control are recognized prerequisites for reliable measurements in some areas of analysis, and modern data processing and management systems are indispensible or are fast becoming so for almost every analytical technique.

Specialized apparatus and equipment are rapidly displacing general purpose equipment so that certain kinds of analytical measurements are limited to laboratories where they are available. In some cases, apparatus dedicated to a specific analytical problem and even to a specific range of analyte concentration may be needed to minimize contamination and memory effects.

Not only are appropriate facilities and equipment required, but they must be adequately maintained. Prescribed maintenance schedules, calibration intervals, and good housekeeping practices must be established and followed to ensure the quality of the work.

With such dependency on facilities and equipment, their availability and use can engender unmerited and even false security to both the producer and user of analytical data. It is a fair statement that suitable facilities and equipment are necessary but certainly not sufficient for the production of quality data. They must be backed up by the other components of a quality assurance system. The complex nature of the problems that can be solved by such facilities and equipment increases, rather than decreases, the need for quality assurance of the measurement process.

GOOD LABORATORY AND GOOD MEASUREMENT PRACTICES

Good laboratory practices (GLPs) and good measurement practices (GMPs) embrace the total experience of analytical chemists in the making of good measurements. GLPs refer to the general practices that relate to many if not all of the measurements made in a laboratory. They are virtually independent of the techniques used and address such subjects as maintenance of facilities, records, sample management and handling, reagent control, and cleaning of laboratory glassware. Good measurement practices are essentially technique specific. They can be extensions of GLPs to the requirements of specific measurement techniques, or they may be completely independent.

The subjects for GLPs and GMPs are best identified and the contents best developed by the laboratory personnel on the basis of their experience and that of their peers. However, the detailed contents will ordinarily be specific for each laboratory due to the special conditions that will be prevalent. GLPs and GMPs must be documented and formally implemented in a laboratory if good measurements are to be expected. Too often, many GLPs and GMPs are assumed to be common knowledge and hence unnecessary subjects for documentation. It is true that many of the things done in a laboratory do not need to be formalized. Indeed, it would be virtually impossible to document every step and every detail of every operation that may be carried out. However, critical operations can be identified, and they should be optimized. Any operation or suboperation that is discovered to be an assignable cause of significant variance or bias is a prime subject for optimization and documentation.

GLPs achieve credibility when they are developed in an interdisciplinary approach. An interdisciplinary committee, for example, should identify the critical practices or operations that affect a large percentage—if not all—of a laboratory's outputs and draft the GLP documents that are appropriate.

GMPs, because of their specificity, should be developed by the personnel using the specific techniques following the same approach as used in the case of GLPs.

In either case, the developers should consider the factors that could affect the precision and those that influence bias when selecting subjects for the documents and developing their contents. Such factors may be classified as those inherent in the methodology and those resulting from how it is used. The first can be improved only by research and development, while the second kind can be improved and controlled, up to a point, by improved laboratory practices. Once developed, GLPs and GMPs should be followed, scrupulously, in order to evaluate them. They should be reexamined periodically for their credibility and amended as necessary.

STANDARD OPERATIONS PROCEDURES

Standard operations procedures (SOPs) describe the way specific operations and methods are to be performed. These include sampling operations, sample preparation, calibrations, measurement procedures, and any operation that is done on a repetitive basis. " Standard " means that it specifies the way the operation is to be done on each occasion, which may or may not be a procedure developed by a standards-writing organization. However, when such is available, laboratories are well advised to consider them since they represent peer judgment and can provide a basis for comparability of data among user laboratories.

When methods are used repeatedly, there is incentive to write them up in a formal manner, but those used on an infrequent basis are often based on an analyst's "know how." Such methodology is seldom optimized and typically may be used differently on each occasion of use. The practice of using unwritten methods should be discouraged. Documentation of what was done is almost always needed. When the writing is after the fact, it is likely to be brief, sketchy, and possibly inaccurate. Describing a method before its use is part of planning and should be done with as much care as that used for all other aspects of a chemical analysis. Colleague review of all methodology is a minimum requirement to ensure understandability and to enhance continuity of a measurement process, as well as to receive the benefit of technical criticism. The writing need not be difficult and time consuming if a standard format is used. This also provides a mechanism for screening for possible omissions in the descriptions and for intercomparison of the various procedures used in a laboratory.

While the use of SOPs may provide a continuity of measurement experience, no methodology should be used blindly. Its appropriateness should be reconsidered at each use. If used infrequently, it may be necessary for an analyst to make a sufficient number of preliminary measurements to demonstrate attainment of statistical control of the measurement process on each occasion.

DISTINCTION BETWEEN GLPs, GMPs, AND SOPs

The distinction between GLPs, GMPs, and SOPs is not sharp. The difference is largely a matter of specificity. In general, a practice provides guidance while a procedure presents direction. Another distinction is that practices rarely, if ever, produce measurement results, although they often affect such results. The scope of a practice typically is broad and is applicable to a number of related situations or operations. GMPs are more specific than GLPs in that they often tailor the latter to specific measurement techniques. Practices are usually written in the indicative mode while procedures should always be written in the imperative mode.

A procedure typically produces a specific result, either in the form of a measurement result or an end product. While every detail cannot be described, the objective of an SOP is to eliminate differences by which the result or end product is obtained. A good SOP should specify the permissible tolerances for control of critical parameters.

While the differences between practices and procedures should be recognized, these should not become barriers to their development and utilization. One should keep in mind the objectives they are designed to reach and use them in a constructive manner.

Table 12.2. Protocols for Specific Purposes*

What they are

- Protocols to define what is to be done in a specific service/project/task/program
- May concern recurring services
- May be tailored to a specific service; approval by management/client desirable in such cases

How prepared

- By responsible authority

Content

- Designation of principal investigator
- Definition of problem
- Specification of model
- Specification of sample(s)
- Data quality requirements
- Specification of data base
- Specification of methodology
- Reference to GLPs, GMPs, SOPs, other PSPs as appropriate
- Specification of controls
- Quality assessment procedure
- Release of data
- Quality assurance responsibilities

*Sometimes called project Q A plans

PROTOCOLS FOR SPECIFIC PURPOSES

The PSP is the most specific aspect of a quality assurance program. It defines explicitly what is to be done in a given measurement situation. The situation may be as small as a simple analytical service offered by a laboratory, or it may be as large as an international monitoring program. The scope of a PSP is outlined in Table 12.2. The degree of detail in any particular PSP will depend on the relative magnitude and importance of the subject addressed in the specific situation.

The PSP is prepared by the authority responsible (or designee) for the measurement operation, sometimes called the principal investigator. This may be the laboratory director or supervisor in the case of work to be done entirely within a single organization, or the principal official of a monitoring program. When a PSP is developed for a particular recurring problem, it becomes the way the pertinent data will be obtained in each instance of use. When data are obtained for an external purpose, all interested parties may need to be involved in developing and/or approving the protocol. Needless to say, input and/or review should be sought at the bench level to make the requirements of the protocol meaningful and applicable.

The PSP is both a planning and an operational document, and each aspect is a worthy reason for its careful preparation. It may be very brief with references to other documents as possible (and this is preferable), or it may be very detailed for major programs for which little precedent exists. The subjects to be addressed in the content are outlined in the following.

Content of a PSP

Designation of Principal Investigator

The person(s) with overall responsibility for implementation of the work should be designated, together with the responsibilities that are delegated. A quality assurance officer also should be designated, as appropriate, together with the accompanying responsibilities. Other subordinate investigators and/or officials should be designated as required, and their duties should be specified by an organizational chart, for example.

Definition of Problem

The problem to be solved should be clearly defined. This should include a concise statement of the objective and scope (limits) of the work to be undertaken. When part of a larger program, an overview of all of the activities may be included with special reference to related ones not in the scope of the present project.

Specification of the Model

This includes the hypotheses that are to be tested and the assumptions that have been made. The general data requirements should be stated together with the precision and accuracy that should be attained.

Specification of Samples

This section should include details as to the number and kind of samples (including location, time, etc.) as required by the model. The methods of sampling, storage, preservation, subsampling, preparation for analysis, and other pertinent information should be prescribed. The manner of distribution and the chain of custody procedures may need to be specified in some cases.

Specification of Data Base

This section should specify the data that should be obtained including that for analytical samples and for quality assessment. Measurement sequences also may be specified. The specification of the data base required to qualify participants (or contractors) as peers in a monitoring program, prior to their participation, is a part of this section [138].

Specification of Methodology

The measurement methods to be used should be specified in detail (SOPs) so that no misunderstanding should arise as to how measurements are to be made. Even when standard methods are specified, additional information may be needed, such as explanatory notes. Any GLPs and GMPs not given in the SOPs but deemed necessary to obtain data of requisite quality must be defined. Reference to appropriate GLPs and GMPs described elsewhere may be sufficient, but they should be readily available and reviewed for their adequacy in the present situation. Likewise, where deviations and/or exceptions from standard methodology are permissible, they should be clearly indicated.

Specification of Controls

This section should describe all activities that will be undertaken to assure the quality of the data. Any necessary preliminary activities such as instrument warm-up, and performance checks should be included, if not part of the SOPs. The corrective actions to be taken and the procedure to be used to indicate when these are necessary should be prescribed.

The control charts that should be maintained, and the rationale for their interpretation should be specified. This will include those to be maintained by the measurement laboratory and those by others, such as a reference laboratory. All control charts should be kept as close to real time as possible so that deviations from statistical control can be detected quickly.

The control samples to be used such as replicates, internal reference materials, Standard Reference Materials, and blinds and double blinds should be specified. In doing so, due consideration should be given to the model (to ensure their appropriateness) and to the data base (to ensure adequacy) of the controls.

Release of Data

This section should specify the requirements that the data must meet before they may be released by the originating laboratory and the criteria that the receiving organization will use to accept data. The persons authorized to judge the above should be designated. When sensitive data are involved, the chain of official release should be defined. Guidelines for release and reporting of data are given in Chapter 22.

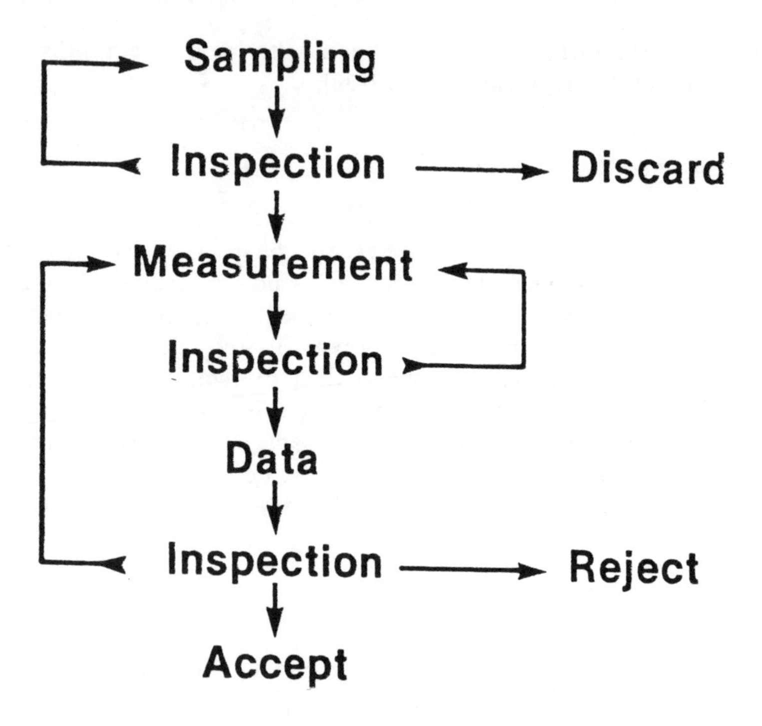

Figure 12.1. Quality control by inspection.

OTHER QUALITY CONTROL TECHNIQUES

Inspection

Inspection is used widely in industrial operations as a quality assurance procedure. Because of its nonquantitative nature, it is most useful for quality control in the measurement process and less useful for quality assessment purposes.

Inspection is largely a subjective examination for malfunction or for the detection of abnormal conditions. A good measurement process will usually have several places where inspection may be used beneficially as suggested in Figure 12.1. Samples may be inspected with respect to preestablished criteria for missing or conflicting information about the sampling process which could be cause for rejection. Also, the samples may need to meet requirements for physical appearance or phase homogeneity, for example. The inspection can take place during sampling, transit, at the time of receipt by the laboratory, and at the time of measurement. The necessity for well-considered criteria to ensure consistent inspection should be obvious.

The measurement process can have performance specifications for the instrumentation (e.g., stability criteria), the laboratory environmental control, and for validation of calibrations before acceptance of data is permissible. Data checks may be made for reasonableness and for expression in proper units, for example. The familiar ionic balance check, long used in water analysis, is an example of a data check.

Inspections do not control quality but may prevent the release and reporting of questionable data. Inspections can be total or conducted on a random basis. A good inspection program, conducted in real time, may make it possible to obtain replacement samples and to make additional measurements as necessary, and thus reduce gaps in data that might otherwise occur.

Documentation

Proper documentation is an essential part of every measurement program. Data must be both technically sound and defensible. While technical soundness is a prerequisite, inadequate documentation can cast suspicion on the technical merits and limit defensibility. In addition, good documentation provides an archival or historical function and the basis for future guidance whenever the same or a similar measurement situation should recur. While there are often staff objections to excessive paper work, one can hardly have too much documentation.

Documentation must address all aspects of the measurement process: the model, the plan, the methodology, the calibration, the samples, and the data reduction. All of these need to be supported with sufficient information and tied together in a comprehensive report that shows the interrelation of each phase and summarizes the data, including all of its limitations. Conclusions, when presented, must be fully supported by readily retrievable documentation.

Details of the documentation process are best addressed in the GLPs. Special instructions for documentation of the data output of measurement techniques should be included in the GMPs. The documentation requirements for a specific project or measurement program should be planned as carefully as all other aspects and specified in the PSPs. In doing so, the documentation practices of all participants should be examined critically, not only for adequacy, but also for uniformity or at least for compatibility. The PSPs may need to specify the format in which the data are to be reported. This may involve the development of special reporting forms.

Concerning laboratory records, hardbound notebooks are still the best means to document measurement processes, in the opinion of this author. Despite the fact that many instruments provide graphical, printed, magnetic, or other forms of data outputs, the notebook provides a coordinating mechanism for such outputs and can contain the measurement rationale behind them. Whatever system is used, documentation and recording of data must be done consistently and accurately. Transcription errors are a common problem that can be virtually eliminated by care and attention to detail. Systematic inspection and periodic review of notebooks and similar primary records are recommended to ensure the general quality of their contents. All data should be recorded in ink, and changes or revisions of notebook entries should be made by crossing out the original entry and substituting the new value. The person making the change should initial and date the entry and state the reason for the change. No erasures of records or data are permissible. A system for review of technical reports, including countersigning by higher authority, is recommended. Also, the staff members responsible for the measurements should sign reports to attest to the validity of the technical contents. Records must be safeguarded and securely stored. Some measurement programs have specific requirements that must be met in this regard.

All of this is not meant to disparage the computer as a means to manage records and data. On the contrary, computer-based systems are unexcelled for management of information and facilitating its ready retrieval. Guidelines for establishing and maintaining traditional notebooks and also for the use of computers to facilitate record keeping are given by Kanare [73].

Training

Adequate education and training are prerequisites for reliable measurement capability. The latter can be distinguished from the former by the degree of specificity. The complexity of modern chemical analysis has increased to the point where undergraduate education is no longer sufficient for the training of analytical chemists. Even highly educated analytical chemists find the need from time to time for additional training due to advances in technology and to the introduction of new measurement techniques. On-the-job training is recognized as a continuing activity which includes indoctrination for a laboratory's quality assurance program for new employees and periodic reviews for the continuing staff. A laboratory's quality assurance manual can provide a good basis for quality assurance training. All training should be direct; no serial training should be permitted.

The importance of informal education and training is not always appreciated. Examples include attendance at technical meetings, participation in standardization activities, attendance at instrumentation exhibitions, and visits to other laboratories. Such activities provide opportunities to associate with colleagues and to exchange ideas as well as to receive specific information. Reasonable periods of time to read technical publications and to browse in the library can bolster morale as well as provide educational values. The need for research personnel to renew and update their knowledge is widely recognized, but a more narrow view is often taken in the case of the analytical personnel, especially if they are considered to provide a support function. Laboratories will find that a modest commitment of resources to encourage educational and

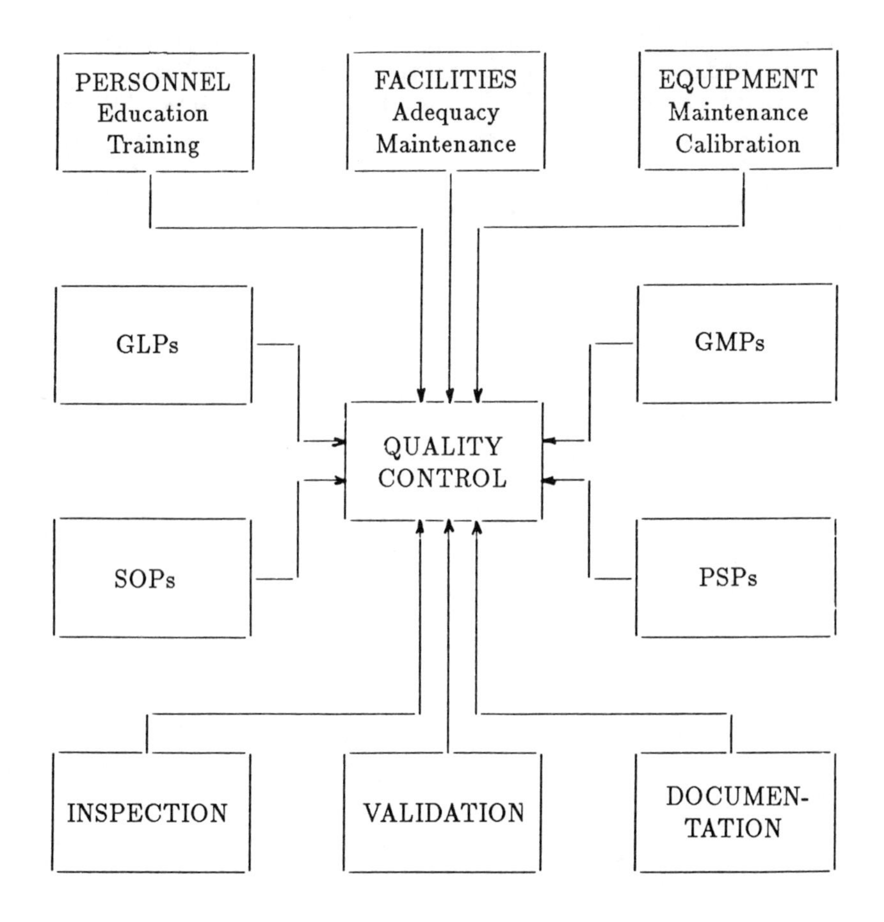

Figure 12.2. Quality control system.

training activities for their analytical staff could produce significant advancement in the state of the art of their measurements and generally contribute to the quality of their data outputs.

THE QUALITY CONTROL SYSTEM

On consideration of the foregoing discussion, it should be obvious that a number of things are necessary, no one of which is sufficient to ensure quality data. In fact, overemphasis on a particular aspect, important as it may be, could be counterproductive in that other contributors to imprecision and bias could be overlooked. A laboratory should recognize that the basic elements should interact and work together in a system as suggested in Figure 12.2. Depending on the nature of the work that is done, the accuracy requirements, the staff and facilities available, and the criticality of all operations, an appropriate and credible quality assurance program should be developed of which quality control is one aspect. The process for doing this is discussed in Chapter 26.

CHAPTER 13

Blank Correction

The degree of control of the analytical blank, i.e., contamination from all sources external to the sample, seriously affects the accuracy of low-level trace determinations. It can also affect the ultimate attainable accuracy in high-accuracy analysis. Most of the sources of the blank are variable, and it is this variability that determines the uncertainty of the blank correction and, often, the lower limit of trace concentration that can be determined with reliability. To improve both the accuracy and lower limit of trace determinations, it is imperative to control the variability of the analytical blank. The only practical way to accomplish this is to reduce the size of the blank itself by controlling the sources of the blank [99].

SOURCES OF BLANKS

The analytical blank results from contamination arising from four principal sources, namely: the environment the analysis is performed in; the reagents used in the analysis; the apparatus used; and the analyst performing the analysis [99].

CONTROL OF BLANKS

The quality control program must give special emphasis to blank control whenever a blank correction is significant. Environmental control can range from simple good housekeeping practices to conducting all operations in an ultraclean room. The latter may be required for all aspects of sample handling up to final introduction of the processed sample into the measurement system. The cleaning and preparation of sample containers also may require clean facilities. A variety of clean benches and enclosures are presently available to provide clean areas in which critical operations can be carried out. They have the advantage of providing easily accessible contamination-free areas at relatively low cost where specific operations can be conducted.

Reagent blanks are derived from all of the chemicals that may contact a sample. Those operations such as dissolutions or extractions, where relatively large quantities of chemicals are involved, can make large contributions to the reagent blank, even if the chemicals used are relatively pure. Water used as a solvent, diluent, or even for washing can be a major source of reagent blank. Both the quality and the quantity of reagents need to be controlled rigorously if reproducible blanks are to be realized. Good practice dictates that the reagents used for a particular set of measurements should come from the same manufacturing lot. Records should be kept of all chemicals used whenever blanks are of consideration.

Blanks can arise from the apparatus used, particularly if chemical operations are involved. Thus beakers, bottles, filters, mortars, sieves, stoppers, and sample lines can contribute both positive and negative blanks. The latter term denotes removals or losses as contrasted to additions (positive blanks) to samples that are analyzed. The control of apparatus blanks involves choosing materials and limiting and controlling the areas and times of contact. Apparatus equipment and facilities dedicated exclusively to a certain kind of sample and even to a narrow range of concentration of analyte may be necessary in some cases. Procedures for cleaning and/or conditioning the apparatus, storage bottles, sample containers, and transfer lines should be developed, written as GLPs, and strictly followed. If the above items are to be cleaned in a central location, special care must be exercised to preidentify the use to be made of specific items and the cleaning schedule to be followed, including the storage after cleaning. Otherwise all items going through the cleaning facility must be cleaned according to the method applicable to the most critical use.

Physical contact of the analyst, including the garments worn, can produce blanks. The analyst must consider how such influences can happen and govern his actions accordingly. In a multidiscipline laboratory, the procedures followed should be uniform and appropriate for the most critical situation.

EVALUATION OF BLANKS

Appropriate control charts provide the best means to evaluate the stability and variability of the blank. Until statistical control is realized, one cannot say anything about the level of a blank. Once it is attained, the control chart or system blank is the best correction to be made to all measurements.

Blank control charts should be analyzed in the same manner as any other control chart. One should look for systematic trends and outliers. If good records are kept, it might be possible to correlate abnormalities with other experimental information to discover assignable causes and corrective measures necessary to obtain acceptable blanks.

STATISTICAL TREATMENT OF THE BLANK

Ordinarily, a measurement is made with all constituents present except the sample, and the measured value is considered to be the reagent (sometimes called chemical) blank. This is subtracted from the value measured for a sample to obtain the "true" concentration of the sample. It is obvious that the blank measurement must be properly made so that the resulting corrections are meaningful.

The following reviews the statistical treatment for evaluation of the blank correction [136].

Let $\bar{C}_m$ = mean of m measurements of the concentration of the measurand in the sample, with standard deviation s_m

$\bar{C}_b$ = mean of b measurements of the concentration of the measurand in the blank, with standard deviation s_b

$\bar{C}_s$ = best estimate of the concentration of the measurand in the sample, corrected for the blank

The statistical uncertainty of $\bar{C}_m$ is given by:

$$\pm \frac{ts_m}{\sqrt{m}}$$

where $t = t_{1-\alpha/2}$ is the value for m−1 degrees of freedom for the 100(1−α)% confidence level.

Likewise the statistical uncertainty of $\bar{C}_b$ is given by:

$$\pm \frac{ts_b}{\sqrt{b}}$$

The uncertainty of $\bar{C}_s$ is obtained by quadrature to give:

$$\bar{C}_s \pm ts_s = (\bar{C}_m - \bar{C}_b) \pm t \sqrt{[\frac{s^2_m}{m} + \frac{s^2_b}{b}]} \tag{1}$$

where $t = t_{1-\alpha/2}$ is the t value for f* effective degrees of freedom (see equation 3) at the 100(1−α)% confidence level.

In the case where the measurement system is demonstrated to be in a state of statistical control and the respective standard deviations are known, equation (1) becomes:

$$\bar{C}_s \pm Z\sigma_s = (\bar{C}_m - \bar{C}_b) \pm Z \sqrt{[\frac{s^2_m}{m} + \frac{s^2_b}{b}]} \tag{1a}$$

where $Z = Z_{1-\alpha/2}$ is 1.96 for the 95 percent confidence interval ($\alpha = 0.05$).

A special case exists when $\bar{C}_m \approx \bar{C}_b$. In this case, the estimate of the standard deviation $s_b \approx s_m = s$, so that:

$$\bar{C}_s = (\bar{C}_m - \bar{C}_b) \pm ts\sqrt{\frac{m + b}{mb}} \tag{2}$$

where $t = t_{1-\alpha/2}$ is the t value for m + b − 2 degrees of freedom for the 100(1 − α)% confidence level and

$$s = \sqrt{\left[\frac{(m-1)\, s^2_m + (b-1)\, s^2_b}{m + b - 2}\right]}$$

In the case of statistical control with $\sigma_m = \sigma_b = \sigma$, one may use:

$$\bar{C}_s = (\bar{C}_m - \bar{C}_b) \pm Z_{1-\alpha/2\sigma}\sqrt{\frac{m + b}{mb}} \tag{2a}$$

The expressions (2) and (2a) are based on measurement of the blank and sample by the same method and apply even if the measurand is not detected in the blank. If the blank and sample are measured by different methods, then equations (1) and (1a) apply. Appropriate values of t, based upon the effective degrees of freedom, f*, must be used in equation (1). These may be computed from equation (2).

$$f^* = \left[\frac{(V_m + V_b)^2}{\frac{V^2_m}{m + 1} + \frac{V^2_b}{b + 1}}\right] - 2 \tag{3}$$

where the variance, V, signifies s^2.

SIGNIFICANCE

Whenever the blank correction becomes significant, it is necessary to measure it with sufficient care. It is clear from the above that blank measurements may need to be made with the same amount of effort as the sample itself, as $C_m \rightarrow C_b$. This fact is often overlooked by experimenters who may make a limited number of measurements of the blank while devoting most of their effort to measurement of the sample.

Blank corrections become increasingly important in the case of measurements close to the limit of detection. The effect of small variability of the blank is magnified in this case. Likewise, even small constant blanks can result in the differencing of two quantities approaching each other in magnitude.

The question of acceptable limits for the blank will now be addressed. The absolute value of the blank would appear less important than its accurate evaluation. However, it is a necessary correction, and good measurement practice dictates that it should be kept within reasonable limits. An empirical rule in the case of trace analysis is to limit the blank correction to no more than ten times the acceptable limit of error for the measurement (preferably considerably less) and, furthermore, that it should never exceed the concentration level expected in the sample. The logic behind the first condition is that up to a 10% error in estimation of the blank would cause no serious difficulties. The second condition is to prevent minor errors in the two measured quantities from introducing large errors in the difference, which is the quantity of practical interest.

In the preceding discussion, the significance of the blank was considered on the basis of its contribution as a concentration factor under the final conditions of measurement. Furthermore, the term C_b contains the sum of the contributions from each source of blank. If C_b is excessive, and if several agents are involved, measurements must be carried out in a suitable program to identify each source and the magnitude of its contribution in order to take corrective actions. Obviously, the magnitude of each source of blank and the final conditions of measurement (e.g., final volume of a solution) must be considered in establishing a permissible level for the blank for each reagent used.

In all of the above discussions, it was assumed that the blank measurement simulates the sample measurement process so that the value of C_b is meaningful. In some cases, it may be difficult or even impossible to fully simulate the sample measurement process unless the sample matrix is present in critical steps of the procedure. If matrix simulation is necessary and cannot be achieved, it may be necessary to independently analyze each reagent for its measurand content and calculate its contribution to the measurement blank.

A related question is the uncertainty of the measured values resulting from uncertainties in the analytical function. Most measurements involve the use of such a function to relate the measured quantity (signal) to the concentration of the sample. Uncertainties in this function must be considered as a measurement uncertainty. The uncertainty in the analytical function is not a significant consideration in the blank correction, provided both measurements use the same function. However, it must be considered in evaluating the final measured value.

There is no way to correct for any blank other than the so-called reagent blank. Rather, one should look for assignable causes and eliminate or minimize them to the extent possible. Ordinarily, such sources are variable and provide unpredictable effects. Good records are essential if they are to be discovered.

While the foregoing discussion has been confined to the analytical blank, the same principles apply to any situation where the difference of two measured quantities is of interest, such as the subtraction of background or control levels from a measured environmental level to determine whether contamination may exist. Obviously, the background or control samples must be measured with sufficient care and in sufficient number to permit reliable decisions on the significance of any observed differences. The empirical N - N′ - N″ rules (for example, 7 - 7 - 7, or 10 - 10 - 10) for field blanks - field samples - spiked blanks are based on such considerations [3]. The interested reader is referred to Chapter 8 and the sampling literature for further discussion on this subject [79].

CHAPTER 14

Control Charts

Control charts are basic tools for quality assurance. They provide a graphical means to demonstrate statistical control, monitor a measurement process, diagnose measurement problems, document measurement uncertainty, and generally aid in methodology development. They may also be used to monitor and/or document critical aspects of samples and sampling operations.

The concept of control charts was first developed in 1934 by Walter Shewhart [121] to describe the outputs of manufacturing processes. Cameron [109] and coworkers at NBS pioneered the adaptation of control charts in monitoring measurement processes. Over the years, a number of modifications of the Shewhart format have been proposed and used. However, this author believes the original format to be the most useful for chemical measurements, and this format will be described in the following section.

FORMAT

There are basically two kinds of control charts. A "property" chart utilizes a single measurement (X chart) or the mean of several measurements ($\bar{X}$ chart) of a selected property of a stable control sample (sometimes called a check standard). An example of an X chart is shown in Figure 14.1. A precision chart utilizes estimates of the standard deviation of the measurement process. In either case, the variable selected is measured occasionally, and the result is plotted sequentially or as a function of time. The first mode has the advantage of coordinate compression while the latter can indicate rate of change of the variable when it occurs. Limits are indicated within which the plotted values may be expected to lie while the process is in a state of statistical control. The limits in Figure 14.1 are simply "3-sigma" limits, following the original Shewhart format, that indicate the bounds within which substantially all of the data should lie when the system is in a state of statistical control.

An $\bar{X}$ chart (Figure 14.2) is preferable because it is less sensitive to blunders (occasional wild results), but such charts may not be feasible for the time-consuming measurements that are the hallmarks of analytical chemistry. Another reason for preference is that means are more likely to be normally

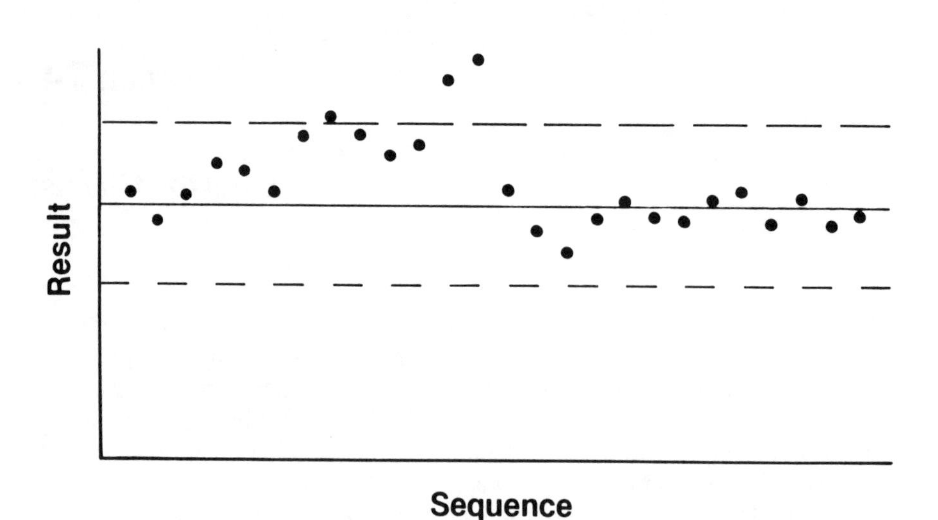

Figure 14.1. Typical X control chart.

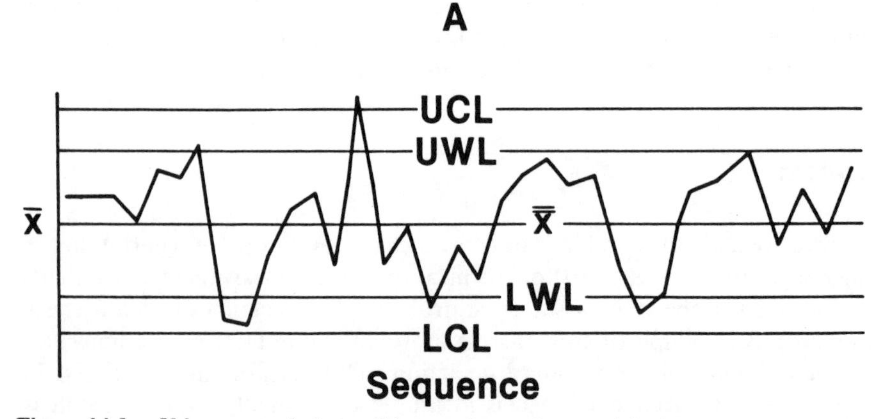

Figure 14.2. X bar control chart with warning and control limits.

distributed than individual values, which is important because a normal distribution is ordinarily assumed when setting control limits. While an X chart has the advantage of requiring less work, additional measurements may be required on occasion to confirm an apparent indication of "out of control."

A precision chart consists of the standard deviation (or the range which is related to it), evaluated at various times, plotted in a fashion similar to that above. Ordinarily, several (at least 4) measurements are needed to evaluate the standard deviation on each occasion, and this requirement is its major disadvantage. The range chart is a useful type of precision chart and will be described later.

The maintenance of property charts and precision charts in parallel has considerable diagnostic value. By intercomparison of the charts, one can usually determine whether bias, imprecision, or both are affecting a measurement process. For example, one would expect bias to be responsible when the property chart indicates out-of-control but the precision chart does not. To facilitate interpretation, the parallel charts should have identical abscissas.

A good arrangement, when feasible, consists of making duplicate measurements of a control sample. This permits the maintenance of an $\bar{X}$ chart for the mean of the measurements and a precision chart based on the range of the measurements as described later.

CONTROL LIMITS

A control chart utilizes a "central line" to define the best estimate of the variable plotted. Around this will be located limits within which a measured value may be expected to lie with a selected probability. Control limits (3-sigma) define the bounds of virtually all values produced by a system in statistical control. Modern control charts often have additional limits called warning limits (2-sigma) within which most (95%) of the values should lie. Only a few should lie between these two sets of limits. When considering control limits, remember that the population standard deviation, sigma, is never known, but only an estimate based on limited data. No practical problems are caused by assuming the estimate to be sigma when limits are set, except in the case of very small data sets, as will be mentioned later.

Control limits can be based on established limits or experimentally established ones. For manufactured goods, the limits could be based on acceptability of product with the central line as a desired value and the control limits as permissible tolerances. Similarly, in measurement, the certified value of a control sample could serve as the central line while the limits could reflect a permissible uncertainty for the measurement process. The above approach is no longer considered favorably, since it does not really monitor either a production process or a measurement process, but merely indicates what can be "gotten away with." This is what the modern concept of quality aims to eliminate—namely, the transition from what is acceptable to what is achievable.

The recommended procedure, almost universally followed today, uses the means of a number of measured values of the variable as the central line and the experimentally estimated standard deviation to establish the control and warning limits.

When a known reference sample is used as the control sample, its certified value may be used as the central line, provided that it does not differ significantly from the measured value and the laboratory's statistics of measurement used to define the control limits. The basis for assignment of such limits is described more fully in the ASTM Manual [7] and summarized in Table 14.1.

Control limits for several kinds of control charts are given in Table 14.1. In the table, s_b denotes the long-term standard deviation—that is, the expected variability when measurements are made on various occasions. The basis for assignment of limits is reasonable, provided s_b is based on at least 14 degrees of freedom. For values based on less information, the appropriate value for t should be substituted for the Z-factors 2 and 3, respectively (see Table C.3).

EVALUATING CONTROL LIMITS

As already stated, reliable estimates of the mean and the long-term standard deviation are required to establish control limits. No less than 7 and preferably 15 independent measurements are needed to obtain initial estimates of these statistics. The measurements are made preferably on different days, but no less frequently than at half-day intervals, so that essentially long-term values can be estimated.

Once initial limits are set, a chart can be established and used (see Figure 14.3). As the chart is used, it will become apparent whether the original limits are realistic. Ordinarily, as much additional data will be required as was in the earlier data base to decide whether the new statistics differ significantly from the original ones, requiring that new limits be set, or whether all accumulated data should be pooled to obtain limits based on an increased number of degrees of freedom. The figure shows two readjustments of control limits around a stable central line.

Table 14.1. Control Chart Limits

X Chart	
Central line	$\bar{X}$ or known value
W L	$\pm\ 2\ s_b$
C L	$\pm\ 3\ s_b$
$\bar{X}$ Chart	
Central line	$\bar{\bar{X}}$ or known value
W L	$\pm\ 2\ s_b/\sqrt{n}$
C L	$\pm\ 3\ s_b/\sqrt{n}$
R Chart (duplicates)	
Central line	$\bar{R}$
U W L	$2.512\ \bar{R}$
U C L	$3.267\ \bar{R}$
L C L = L W L	0

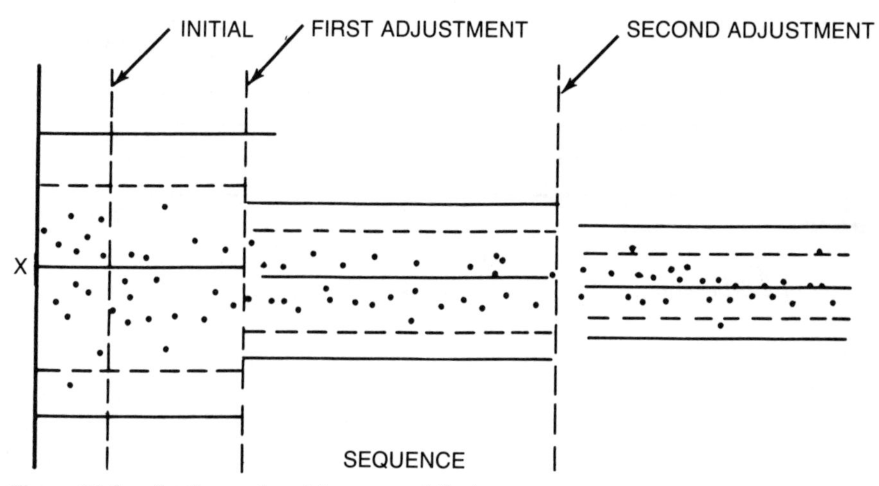

Figure 14.3. Setting and revising control limits.

CONTROL SAMPLES

The control samples must have a high degree of similarity to the actual samples analyzed. Otherwise, one cannot draw reliable conclusions about the performance of the measurement system on test samples from its measurement of control samples. Obviously, the control samples must be sufficiently homogeneous and stable so that individual increments measured at various times will have less variability than that of the measurement process.

When the value of the property measured is known with sufficient accuracy, both the precision of measurement and any systematic measurement errors (bias) can be estimated. Even if the exact composition is not known, a suitable control sample can evaluate the stability and precision of the measurement process.

Ideally, a control chart should be established for each kind of measurement that is made, and indeed, for each parameter level as well. This becomes infeasible and perhaps impossible in any but very routine measurements on very well-defined materials, such as might be encountered in a limited scope manufacturing process. Accordingly, one must look for typical control samples for critical measurement situations. The duplicate sample control chart, to be discussed later, is a partial solution to this problem.

FREQUENCY OF CONTROL MEASUREMENTS

The required frequency of measurement of control samples will depend on a number of factors, first of which is the known stability of the measurement process. A very stable process will require only occasional monitoring, but one

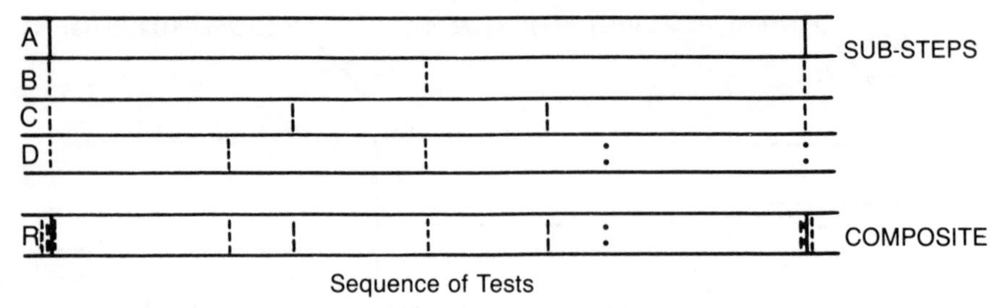

Figure 14.4. Length of run concept for frequency of measuring control samples.

must be sure that one's optimism in this respect is well founded. The importance of the decisions based on the measurement data is another factor. Control intervals should be chosen with the realization that any data obtained during the period from last-known-in-control to first-known-out-of-control will be in jeopardy. Another factor is related to the cost in time and effort of measuring a control sample. If this is relatively small, one is well-advised to err on the conservative side and to measure more controls than the minimum number.

Guidance in choosing control sample intervals can be found by applying the "length of run" concept. For example, the interval between retoolings in a production process may be considered as a production run. Each run could conceivably produce a somewhat different population of products that should be evaluated separately. Measurement systems may be studied for possible runs (see Figure 14.4) such as:

- between calibrations
- between days
- between shifts
- between rest periods
- between periods of use
- between critical adjustments

The magnitude of possible changes between runs could be investigated and significant ones used to establish the length of a significant run. Incidentally, the same concept may be used when troubleshooting to investigate the cause of a significant change in performance of a measurement system.

When a significant run is identified, control samples should be measured within them. At least two measurements, one at the beginning and one at the end of the run, are recommended. An additional one at midperiod is highly desirable.

In the absence of definitive information, arbitrary schedules may be used for the measurement of control samples. These may be based on the experience of the laboratory and sometimes are prescribed by users of the data, such as regulatory agencies. Table 14.2 and Table 14.3 are examples of arbitrary

Table 14.2. Quality Assessment Schedule Using Internal QA Samples

	Calibration—full expected range
*	IQA_0
	Test samples—group 1
*	IQA_1
	Test samples—group 2
	.
	.
	.
	Test samples—group n–1
*	IQA_{n-1}
	Test samples—group n
*	IQA_n
*	Calibration mid-point

Notes

*Decision point

1. Maintain control charts as follows
   X control chart, IQA
2. System must be in control at decision points
3. At least 2 groups with a maximum of 10 samples in each group
4. IQA_i indicates the occasion that the same internal reference sample is measured

Table 14.3. Quality Assessment Schedule Using Duplicates or Split Samples (D/S)

	Full Calibration
*	Calibration check—midrange
	Sample 1
*	Sample 1 (D/S)
	Samples 2–9
	Sample 10
*	Sample 10 (D/S)
	Samples 11–19
	Sample 20
*	Sample 20 (D/S)
*	Calibration check—midrange
*	Calibration check—midrange (duplicate)

* Decision point

1. Maintain R control charts
   a. Duplicate midrange calibration point
   b. Duplicate/split sample
2. System must be in control at each decision point
3. If more than 20 samples, repeat sequence
4. If less than 20 samples, divide into two groups

schedules that may be useful in typical measurement programs. The intervals between measurement of control samples should be set conservatively at first and then may be increased or decreased, depending on the observed stability of the measurement process.

IDENTIFICATION OF POPULATION

An important issue in the development of a control chart is the identification of a population of measurements for which it may be applicable. Technical judgment is of prime importance when defining the unifying characteristics of diverse measurements that would permit their inclusion in a population having essentially the same precision of measurement. Utilization of the same measurement principle, the same equipment, the same analyst, and the same measurement steps would ordinarily be grounds for consideration for inclusion. Procedures that incorporate similar chemical processing steps also may be considered together in some cases. Within limits, the precision of a process may be independent of the concentration level of an analyte. However, widely differing levels cannot be represented by the same control chart. One must always be cautious when extending inferences from known to unknown measurement systems.

STRATEGY OF USE

Control charts may be used to show that the system monitored is within expected limits, to signal systematic departures, and to identify inconsistencies in precision. The control limits are based on probabilistic considerations, as already described; hence, in-control decisions are statistically supported. When a relatively large amount of control chart data is available, a 1-sigma zone can also be established, and the percentage of points lying within the various zones can be used to judge statistical control. Thus:

- ± 1 sigma should contain $2/3$ of the points
- ± 2 sigma should contain $19/20$ of the points
- ± 3 sigma should contain "all" of the points

when the system is in statistical control.
Other indicators of out-of-control include

- 2 successive points outside of the 2-sigma limits
- 4 successive points outside of the 1-sigma limits
- any systematic trends

An example of a systematic trend is the existence of runs. These can be a series of points proceeding in the same direction (up or down) or points residing on

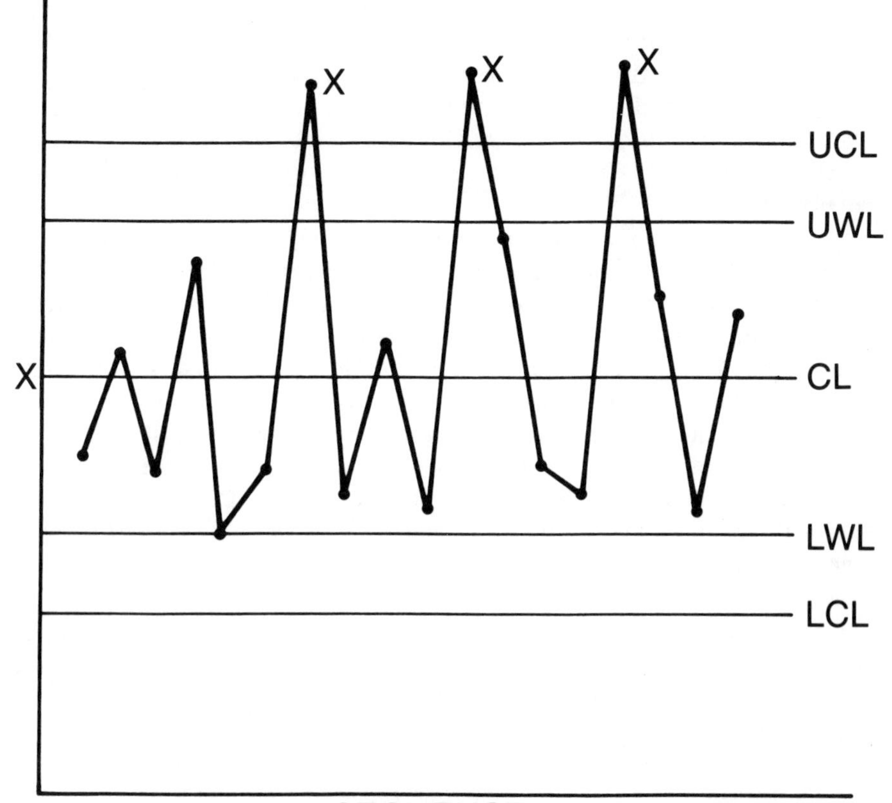

Figure 14.5. Identification of rare events.

the same side of the central line, even though all of them are within control limits. In deciding whether an observed run is unlikely, based on chance, the number seven may be used. Thus, the probability of occurrence of seven consecutive events as described above is only about 1 in 100, based on chance alone. Such behavior can be an early warning of trouble and indicate the need for preventative action.

Referring to Figure 14.1, a systematic trend will be noted in the early data that should have been obvious to the analyst. The chart indicates an upward drift culminating in several points outside of the control limits. The systematic problem would have been even more obvious if the chart had included warning limits and the guidelines recommended above had been followed. This assumes that the control chart had been maintained and interpreted on essentially a real-time basis. Too often, charts are prepared with a considerable time lag so that they provide only a history of performance with little opportunity to control the measurement process.

In addition to looking for systematic trends, one should look for patterns for the occurrence of out-of-control situations. Distinctive marks such as X's may be used to identify suspicious points (see Figure 14.5), and these marks should be scrutinized for patterns, periodicities, or correlations with abnormal operating conditions. If assignable causes can be found, their correction or elimination could enhance the performance of the measurement process [147].

The identification of assignable causes can tax the ingenuity of the investigator. Obviously, a profound understanding of the measurement principle and of the various steps of the procedure is necessary, and complete and careful documentation of all aspects of the measurement process is required. Before undertaking such an investigation, one should be sure that any apparent change in behavior is real and significant. Visual inspection of the control chart may be the best way to identify gross trends consisting of short runs, changes of slope, jumps, and shifts. The significance of apparent differences in means can be evaluated by use of the statistical t test. Differences in precision may be confirmed by the F test. These statistical tests are described in Chapter 4.

Measurement systems used occasionally are more likely to show out-of-control behavior than those used constantly. After disuse, reattainment of statistical control may be slower than anticipated. Also, an analyst may need to make a number of preliminary runs to regain competence in the use of a methodology that had been used very successfully in the past. In fact, demonstration of statistical control is a prerequisite to reliable measurement, and the control chart is a superior technique for accomplishing this.

ASSIGNMENT OF UNCERTAINTY

When a measurement process is demonstrated to be in statistical control, as evidenced by control chart performance, the operational statistical characteristics of the process, documented by the control limits, may be assigned to appropriate measurements produced. Appropriateness should be interpreted on the basis of the similarity of the control samples to the test samples. For example, if the similarity approaches identity in composition and matrix, a confidence interval for the mean of the test results ($ts\sqrt{n}$) can be computed on the basis of the value of the standard deviation used in setting the control limits. The value to be used for t will depend on the number of degrees of freedom on which the control limits are based (and the confidence level desired), while n indicates the number of replicate measurements of the test sample. This is discussed further in Chapter 22.

Table 14.4. Factors for Use in Duplicate Range Charts and Other Sets of Replicates

Number in Set	UWL	UCL	LCL
2	2.512	3.267	0
3	2.050	2.575	0
4	1.855	2.282	0
5	1.743	2.115	0
6	1.669	2.004	0

RANGE CONTROL CHARTS

The problem of appropriateness of control samples can be circumvented by using replicate measurements of the actual test samples to monitor measurement precision. A special case is the duplicate measurement of a reasonable number of test samples. It will be remembered that the range of a set of measurements (in this case, the difference of the duplicate measurements) is related to the standard deviation. Accordingly, range control charts can be developed and used to monitor measurement precision. However, they do not evaluate systematic departures that would affect each measurement in the same way. Thus, instrumental shifts and calibration drifts would go undetected. One would need to include other control samples and/or calibration checks to monitor such problems. Despite this deficiency, range control charts are considered by the author to provide an excellent means to evaluate measurement precision because actual test samples are involved, leaving no unanswered questions about interpretation of the results.

The following discussion considers only charts based on duplicate measurements. However, sets of any size (triplicates, etc.) could be used similarly.

Approach

A range chart utilizes the average value for the range, based on past experience, as the central line and control limits within which a percentage of future values are expected to lie. The average value of the range is calculated from k sets of duplicate measurements by the expression:

$$\bar{R} = (R_1 + R_2 + \ldots + R_k)/k$$

The control limits are multiples of $\bar{R}$ using the factors found in statistical tables [7]. The numerical value for the factor depends on n, the number of measurements in the set, and the confidence level for judgment of control. The factors for use in duplicate range charts (number in set = 2) and for several other sets of replicates are shown in Table 14.4.

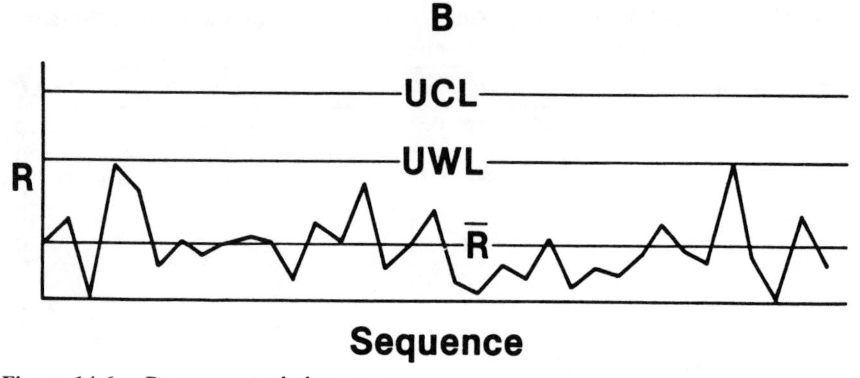

Figure 14.6. Range control chart.

A laboratory may use records of its previous work or conduct a series of measurements to determine $\bar{R}$. As many duplicate measurements as feasible, but no less than 8 and preferably at least 15, should be made to calculate $\bar{R}$. The duplicate measurements preferably should be made on different days, but no more frequently than at half-day intervals.

A typical range control chart is shown in Figure 14.6. When establishing a chart, it is a good idea to plot the values of R sequentially and separate them into groups of four. A horizontal line representing the mean value, $\bar{R}$, is then drawn. The plot should show no systematic tendencies with respect to the central line or between tetrads. Passing this test, the control limit and warning limit lines are drawn, and the chart is ready for use.

As the control chart is maintained, it can become more firmly established. Subsequent plots of Rs should exhibit no systematic tendencies nor should they show any closer clustering around the central line. If they do, it suggests that either the original estimates of $\bar{R}$ were grossly wrong or that something about the measurement process has changed. Judgment with respect to precision is made by calculating original and later values of s from the corresponding $\bar{R}$ values (see Table C.1) and then making an F test. In the case of improved performance, calculate new control limits based on the later data and continue to use the updated control chart. In the case of poorer performance, decide whether the original estimate was too optimistic or whether performance has indeed deteriorated. In the latter case, corrective actions should be taken to restore the earlier performance. If no change has occurred, combine all data to calculate a better value for $\bar{R}$ and new control limits.

It is recommended that at least eight new sets of data be accumulated before any attempt is made to revise control limits. As the values are based on increasing amounts of data, the addition of additional data to the base will be less influential. Accordingly, it is further recommended that no recalculation be made unless gross changes are suspected or the additional data is at least 25% of that already in the base.

Use of Range Control Charts

Duplicate measurements may be made of all of the samples measured, of selected samples in a measurement series (see Table 14.3, for example), of reference materials, or any combination of the above as designated in the laboratory's quality assurance program. When the values of R are plotted on the chart, the following courses of action are recommended:

1. If R is within the warning limits, accept all related data.
2. If R is outside the control limits (UCL), the system is out-of-control. Reject all data since last-known-to-be-in-control and take corrective actions. Reestablish control before accepting data.
3. If R exceeds the UWL but is within the UCL, accept the measurements, tentatively. If the R for the next sample is within the UWL, accept all previous data. If it exceeds the UWL, reject all data taken since the system was last-known-to-be-in-control and take corrective actions. Reestablish control before accepting data.
4. Reestablishment of control should be demonstrated by the results of three consecutive sets of measurements that are in control.

Because the control and warning limits are based on 95% and 99.7% confidence, respectively, a system in statistical control should rarely exceed them. If out-of-control is observed too often, either the limits are not realistic or the system has problems that need correction. Suggestions already made, as well as those contained in Chapter 24, should be considered to improve the system.

Subspan Range Control Charts

The number of control charts to be established can be minimized by using the subspan concept. Every analyst knows that there is no practical difference in the precision of measurement of samples of similar composition. For practical purposes, it is possible to subdivide a span of concentrations into several subspans, as shown in Figure 14.7, each with a precision of measurement that is essentially constant. Thus the span 1–100 might be subdivided into the subspans 1–10, 10–50, and 50–100. In this case, three control charts, one for each subspan, could serve the entire span. The charts should be based on $\bar{R}$ values for samples in the middle of each subspan. For very large spans, a larger number of subspans may be needed. There is the quandary of what to do with data at or near the boundaries of the subspans, but this does not often present a serious problem.

Range Performance Charts

It is possible to maintain a single chart to cover a wide and perhaps the entire span of measurement with only a small sacrifice of utility. The result is a performance chart which only a purist would say is not a control chart. Such a chart is shown in Figure 14.8. Ordinarily, $\bar{R}$ will vary with analyte concentra-

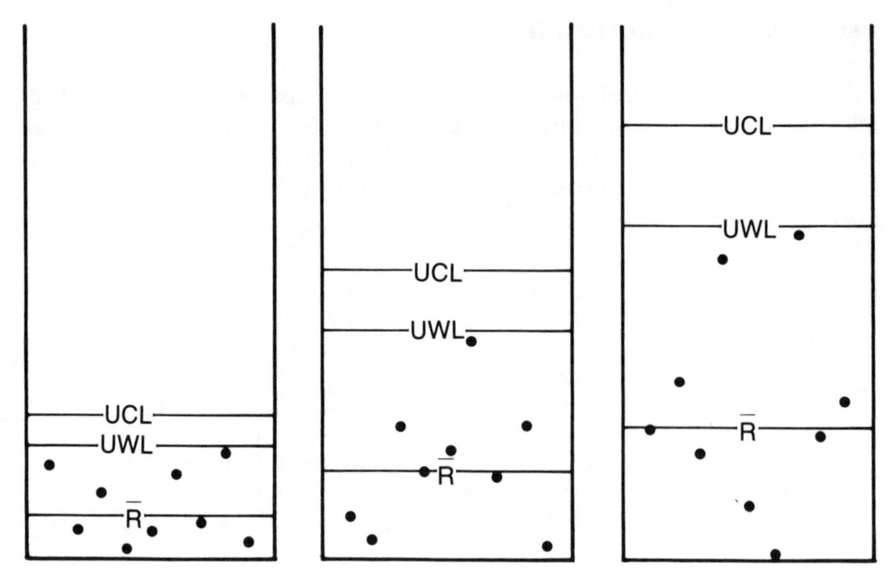

Figure 14.7. Sub-span control charts.

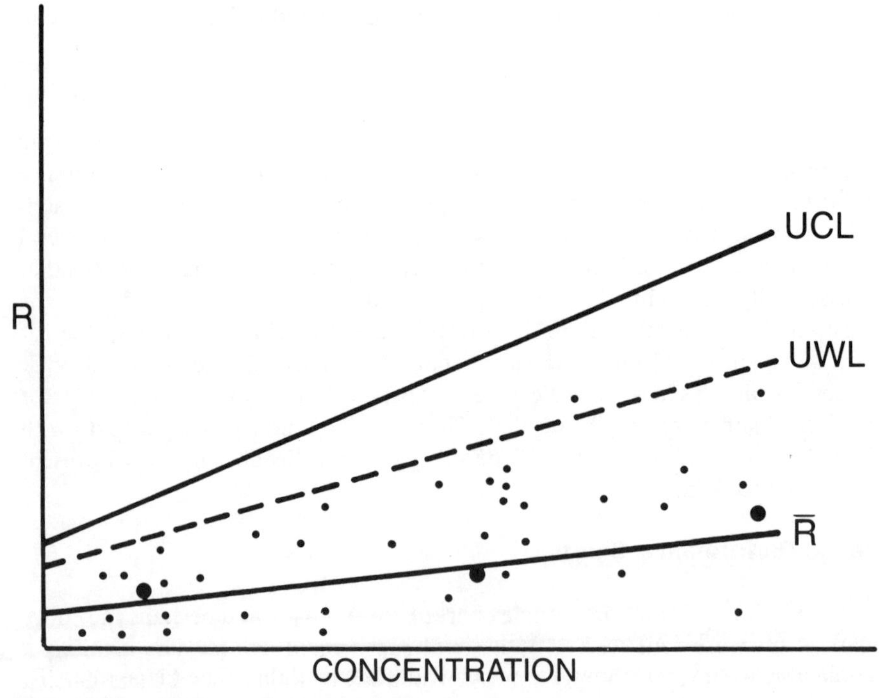

Figure 14.8. Range performance chart.

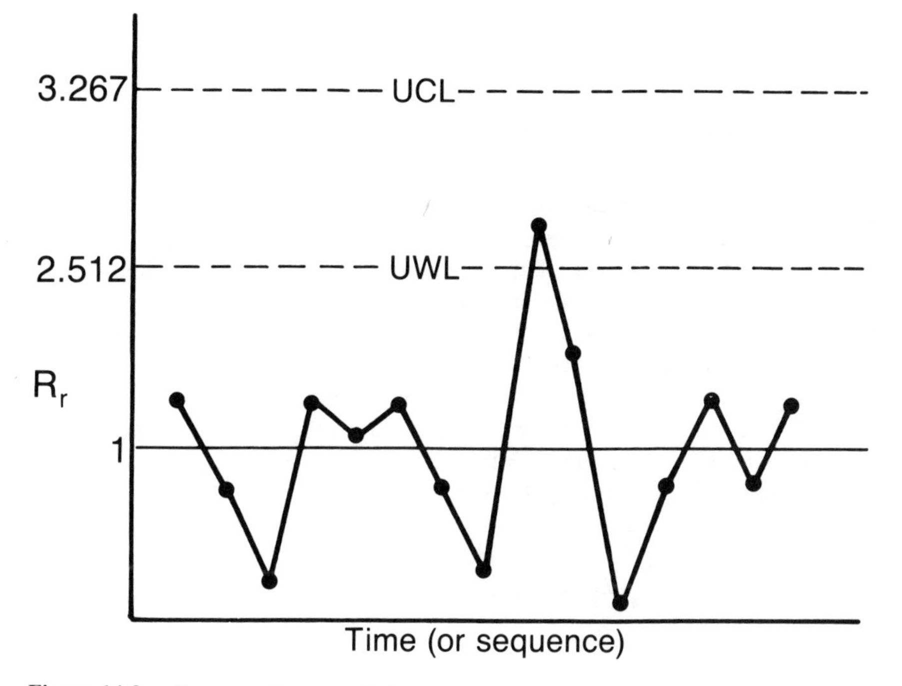

Figure 14.9. Range ratio control chart.

tion level in a linear manner. Values for samples at low, middle, and high level, respectively, are measured and plotted as a function of concentration. A linear fit usually represents the expected value of $\bar{R}$ at each level. The warning and control limits are simply the appropriate multiples of $\bar{R}$. As duplicate samples are measured, the value of R is plotted on the chart, and control is judged on the basis of its being within the limits, as before. The chart is convenient to use despite the fact that the time or sequence axis has been lost. In fact, this is the only reason why such a chart cannot rightly be called a control chart. Color coding or lines may be used to indicate sequence and thus to identify trends. However, the advantage of a single chart to provide so much performance information makes it very attractive in monitoring a measurement process.

Range Ratio Control Charts

With little extra effort, one can construct a range ratio control chart that is applicable to the entire span for which R-data is available. The range ratio, R_r, is defined as:

$$R_r = \frac{R_0}{\bar{R}_c}$$

where R_0 = the observed range and $\bar{R}_c$ is its expected value.

$\bar{R}_c$ may be found from a plot of $\bar{R}$ as a function of concentration. It is clear that, on average, one would expect to obtain a value of 1 for R_r. It can be shown that 95% and 99.7% of the time, respectively, the values should not exceed the values of the factors used to compute UWL and UCL given in the table on page 139. Thus, the UWL for R_r is 2.512 and UCL is 3.267, both for the case of duplicate measurements.

A typical range ratio control chart is shown in Figure 14.9. It is used in the same way as a conventional R control chart. From the observed range and the value of $\bar{R}_c$, R_r is calculated and plotted as a function of its sequence. Decision with respect to in control is made as discussed earlier.

COLLATERAL CONTROL CHARTS

Control samples consisting of internal reference materials (IRM) and Standard Reference Materials (SRM) may be used together in a measurement program and the results intercompared using control charts. Rarely would the two materials be sufficiently similar to permit plotting of the results on the same chart. However, the results could be plotted on parallel charts and intercompared, even if the SRM were measured less frequently than the IRM. Ordinarily, the IRM will be measured more often than the SRM. If control is demonstrated for the IRM and likewise for the SRM at each time of use, traceability could be claimed just as if the SRM had been used every time. The limitations as to what can be claimed to be traceable are the same as in the case of exclusive use of an SRM and are discussed more fully in Chapter 18.

CRITICAL COLLATERAL MEASUREMENT CONTROL CHARTS

Any measurement process that provides critical data or information for use by the main measurement process must be in statistical control, and the best demonstration of this is through the use of control charts. An example is the chemical blank. Quality control procedures should be used to stabilize the blank, which is measured periodically. A control chart of such measured values attests to the stability of the blank and estimates its variability. Ordinarily, the best estimate of the blank to be applied to all measurements is the process blank—namely, the central line of the blank control chart. The control chart will also estimate the standard deviation of the blank to be used in computations of uncertainty, as discussed in Chapter 13.

Other examples of collateral measurement control charts include detection limits and the slope of a calibration curve. Parameters that contribute critically to the measurement process, such as the environmental factors of pressure and temperature, in some cases are also excellent candidates for control chart evaluation.

RECOVERY CONTROL CHARTS

The periodically measured efficiency of recovery of a spike or surrogate is an excellent candidate for control chart evaluation. Depending on the stage at which the spike or surrogate is added, information on the efficiency, stability, and variability of recovery can be evaluated. This becomes much better information than the yes or no decision on attainment of at least a minimum recovery that is sometimes used. Though application of a recovery correction is usually not merited, there are exceptions. In such cases, the control chart recovery is the one that should be used. In all cases, the attainment of reproducible recovery is evidence of statistical control of appropriate parts of a measurement process. Its evaluation by means of a control chart is an important aspect of some quality assurance programs.

OTHER USEFUL CONTROL CHARTS

The imagination of the analyst is perhaps the most severe limitation on the kinds of control charts that could provide valuable analytical information. On the practical side, a cost analysis may need to be made to determine which of them would be cost effective. The following kinds of charts have been beneficial in some laboratories:

- instrument performance factors
- operator performance
- equipment/instrument charts
- sample/sampling parameters
- test weight (if weighing is a critical factor)
- a second buffer (other than that used for calibration of a pH meter)

ADDITIONAL REMARKS

Control charts should be maintained and examined in as close to real time as possible. In this way, systematic trends and out-of-control can be discovered promptly, and suitable corrective actions can be made possible. Such corrective actions can be preventative and greatly improve a measurement process. However, one must remember that hasty reactions should be avoided. Hence, a sound and consistent strategy should be adopted for interpretation and response to control chart indications.

Control charts should be kept at several levels of the supervisory chain. The bench level needs to devote the largest amount of effort to this activity since it has the best opportunity to control a process and to institute corrective actions as necessary. Supervisors can insert control samples into a measurement program at less frequent intervals to monitor the degree of control by the bench. Successively higher levels can monitor with even less frequency, provided that each lower level can show evidence of effective control. The further the monitor is removed from the point of action, the less can be controlled. When monitoring is sufficiently removed in time and/or space, historical performance is all that can be evaluated.

Blinds vs Double Blinds

Blinds describe control samples, the expected values of which are unknown to the analyst. Double blinds describe control samples that cannot even be identified as such by the persons making the measurements. The objectivity of a test is often considered to increase as one uses knowns → blinds → double blinds. It is the opinion of this writer that this issue may be overplayed and that the excessive use of blinds and double blinds can be counterproductive in that an air of distrust can be generated in a laboratory. Knowns are best suited to daily control while double blinds are best suited to performance appraisal. As good control is achieved and demonstrated on a continuing basis, the need for performance appraisal is diminished so that it becomes an occasional confirmatory exercise.

CHAPTER 15

Principles of Quality Assessment

Quality assessment techniques consist of ways in which the measurement process may be monitored in order to infer the quality of the data output. They provide assurance that statistical control has been achieved and is maintained as well as estimates of the accuracy of the data. Table 15.1 lists a number of approaches that may be used. The central theme is the concept of redundancy. Precision can be evaluated only by replication using stable samples of known and even unknown composition. Evaluation of bias requires the repetitive measurement of samples of known properties, although fewer measurements may be required when statistical control of the measurement process is demonstrated by other means. The quality assessment techniques listed in Table 15.1 are classified as internal or external according to the source of the assistance needed to implement the assessment technique.

INTERNAL TECHNIQUES

Repetitive measurement of the samples actually tested is the classical way to evaluate precision. However, this is a time-consuming process if not carefully planned, and it is the objective of other quality assessment techniques to minimize the number of such measurements. Duplicate measurement of an appropriate number of test samples often can provide much of the evaluation of precision that is needed and eliminates all questions of the appropriateness of the quality assurance samples.

Internal test samples may consist of internal reference materials (IRMs), split samples, spiked samples, and surrogates which are measured in suitable test routines to evaluate the precision of the measurement process. Measurements of all such samples, and indeed of all quality assurance samples, are best interpreted using control charts. In this way, a few current measurements can join the experience of the past to demonstrate statistical control and bolster confidence in the measurement process.

Whenever a laboratory believes that a given type of sample or a similar one will be analyzed at a future date, it should consider reserving a portion for remeasurement on that occasion. This practice will give confidence to future

Table 15.1. Quality Assessment Techniques

Internal
Repetitive Measurements
Internal Test Samples
Control Charts
Interchange of Operators
Interchange of Equipment
Independent Measurements
Definitive Method Measurements
Audits
External
Collaborative Tests
Exchange of Samples
External Reference Materials (ERM's)
Standard Reference Materials (SRM's)
Audits

measurements when they agree with previous experience. Furthermore, the additional measurement experience will probably provide increased competence that could narrow the limits on the earlier sample, if that should be of interest. By judicious selection, such samples could become IRMs or approach such status, provided they meet the requirements outlined in Chapter 17.

The preceding discussion has been concerned with the evaluation of the precision of measurement. If the composition of an IRM is known with sufficient accuracy, bias may be evaluated. If not, a laboratory may use other techniques to internally evaluate bias, but these are time consuming.

Measurement bias can be operator-, equipment-, and/or methodology-dependent. Internal approaches that may be used to investigate such bias include interchange of operators and apparatus/equipment, measurement of selected samples by independent techniques, and comparison of measurement results with those obtained using a definitive method.

Most measurement results should be independent of who made them and the apparatus or equipment used, provided adequate calibrations were performed. If not, one should look for inadequacies in the calibration procedure and in the SOP used to make the measurements. Control charts maintained for both operators and equipment should identify any problems in these areas and thus the appropriate corrective action. One must remember that only large differences are significant when nonstatistical judgments are made. As small biases (and especially small differences of precision) are of concern, statistical evaluations based on a relatively large data base are required.

Comparison of results with those obtained using a definitive method is another internal approach to investigate bias. Unfortunately, there are only a limited number of documented definitive methods, so this approach is not often possible. If a definitive method can be used, replicate measurements on the same sample, using it and the method under investigation, will be needed to make a decision on bias, unless the former is very much more precise than the latter. Statistical considerations will provide guidance on how many measurements using each will be required.

In addition to the approaches already described, the agreement of results obtained using the method under investigation and those of an independent method provides some evidence of the absence of bias. However, it must be considered that both methods could be producing biased results for the same reason. This could happen if the two methods had some step in common which produced bias in both sets of results. A common dissolution step or an extraction step would be an example.

EXTERNAL TECHNIQUES

External evidence for the quality of the measurement process is important for several reasons. First, it is the easiest approach in that it can minimize much of the effort required for internal evaluation. Second, it minimizes the danger of error due to introspection. Several procedures may be followed to provide external evidence of the quality of a measurement process. These can confirm the internal evaluation of precision and provide independent assessment of any bias (or lack thereof). Activities include participation in collaborative test exercises, exchange of samples with other laboratories, and the analysis of reference materials obtained from external sources (ERMs). The National Bureau of Standards Standard Reference Materials (SRMs) [134] are unexcelled as test materials to evaluate the measurement process when they are properly used [131]. They have the distinct advantage of wide acceptance, and therefore provide a basis for intercomparison of measurement systems and test data taken under diverse conditions and by various laboratories.

Collaborative test exercises provide the opportunity to compare performance with that of others. If the parameters of the test samples are known with accuracy, bias can be evaluated. Exchange of samples with other laboratories or with other groups within the same laboratory can provide evidence of agreement or disagreement, and this can be used to make inferences about bias or the lack thereof.

The use of reference materials (RMs) to evaluate measurement capability is the procedure of choice when suitable RMs are available. Because of its importance, the rationale of reference material use is discussed separately in Chapter 17. Suffice it to say here that reference materials have the advantage of the ability to test the total measurement process provided there is no question of their appropriateness.

From this discussion, it should be clear that every laboratory can and must have in-house capability to estimate the precision of its measurements. The internal evaluation of bias is difficult, but can be facilitated by the use of externally available techniques. No matter what the source of information, any evidence of malperformance, whether it be lack of precision or intolerable bias, should be carefully investigated, and appropriate corrective actions should be taken as necessary. After such actions, follow-up measurements should be undertaken to verify that the problems have been eliminated.

QUALITATIVE IDENTIFICATION

Assessment of the accuracy of qualitative identification is difficult. The concept of control charts is not readily applicable, and certified reference materials for qualitative identification are not available at detection levels where the problem of identification is most acute. Ordinarily, reliance is based on the known selectivity of the methodology and knowledge of the absence of potential interferents. The knowledge and experience of the analyst can be a key factor in decisions on detection.

Identification using independent techniques and/or several modifications of the same technique (e.g., different columns) is useful for verifying the qualitative identification of a measured analyte. Spot checks by a reference laboratory can be useful both for verification of measured analytes and to ensure confidence in questions of "detect" and "nondetect."

OTHER TECHNIQUES

Audits are an important quality assessment technique that may be conducted internally and externally. The general aspects of audits are discussed in Chapter 19.

Figure 15.1 summarizes the foregoing discussion and shows how all of the techniques mentioned must work together in an integrated system. It also contains two additional techniques not previously introduced. Statistical analysis of data from all sources is an ongoing activity to determine the significance of apparent differences. Every laboratory must have a clear understanding of statistical principles as well as skill in their use. Modern computers can provide the latter, but the former is gained largely by experience. In fact, the misuse of statistics is as fruitless as their disuse. A staff statistician could be one of the most important positions in a modern analytical laboratory.

Another item identified in Figure 15.1 is introspection. This is not a technique, but an attitude and a philosophy. Quality assessment must be both a philosophy and a practice. A good laboratory and all of its staff will have a desire for excellence. Its basic QA program and philosophy will

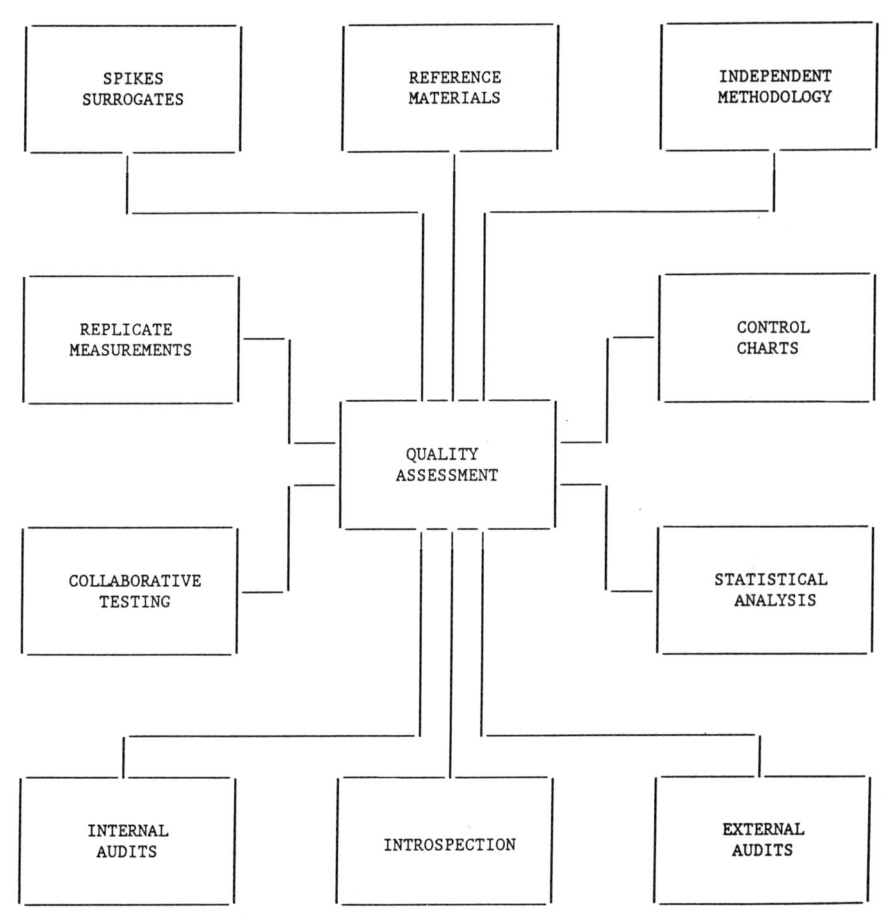

Figure 15.1. The quality assessment system.

provide the incentive to continue surveillance of the quality of its work and to take corrective actions whenever necessary. The specific quality assessment procedures to be followed will be defined in its PSPs.

CHAPTER 16

Evaluation Samples

Any test sample that is used to evaluate any aspect of a measurement process may be called an evaluation sample. Such samples may be used to evaluate or collaboratively test methodology or to test the proficiency of an analyst/ laboratory. They may also be used in a laboratory's quality assessment program. The requirements are generically the same, and the methods of preparation are similar and largely independent, regardless of the use. The specifications may be more stringent in some cases than for others, and these will depend upon the accuracy requirements for the end use of the sample.

PRIME REQUIREMENTS

The first requirement for an evaluation sample is a reasonable match of matrix with that of the usual test samples. In this context, matrix match means equivalency of both the composition of the sample and the kind and level of minor constituents and impurities that could be present. The level of the analyte of interest likewise must be reasonably close to that of a typical test sample.

Evaluation samples must be reasonably homogeneous so that subsamples have an insignificant variance of composition. Otherwise, each evaluation sample must be individually certified.

Stability is a third critical requirement. The samples should exhibit no significant deterioration or change in analyte level during the expected period of use. A shelf-life determination may be necessary to confirm compliance with the stability requirement. The procedure described to determine holding time (see page 63) may be used for this purpose. The objective, in this case, would be to determine that no significant deterioration occurs over a specified service period for the reference material.

TYPES OF EVALUATION SAMPLES

The reference material, and particularly a Certified Reference Material (or SRM), is the highest class of evaluation sample. Such materials are ordinarily certified on the basis of accuracy for the most critical anticipated use, but the expressed certification limits should be consulted when the use of any specific SRM is considered. Because of their special role in measurement, they are discussed more fully in Chapter 17.

Stable test samples, accumulated by a laboratory, can find excellent usage for monitoring the stability of a measurement process. They can be used to evaluate precision and also bias when an accurate value can be assigned for the analyte. Care should be exercised in using such materials so they do not deteriorate or become contaminated by prolonged use. Heterogeneous materials may stratify on standing and need to be reconstituted before every use to avoid possible magnification of heterogeneity problems.

Split samples refer to the procedure in which subportions of well-mixed samples may be used to evaluate a measurement process. The variance of such subsamples should be negligible so that variations in their measured results reflect only measurement variance. The critical need to carefully mix the main portion is evident if subportions are to achieve a requisite degree of homogeneity.

When evaluation samples cannot be obtained by the approaches described above, spikes and/or surrogates may be prepared. The only difference is that spiking refers to the addition of the analyte to be measured to a sample or suitable matrix, while a surrogate is a different substance which has the same measurement problems as the analyte of interest. When this latter approach is used, the surrogate should be chosen with care. Labeled analytes, isotopes, and isomers are the best surrogates. If there is any question of matrix effects, the additions must be made to the matrix analyzed.

The chief objection to spikes or surrogates is concerned with whether an artificially added constituent can be recovered with the same efficiency as a naturally incorporated constituent. Experimental evidence may be needed to answer this question.

The preparation of evaluation samples by synthesis must be done with considerable care. Gravimetric addition is the method of choice because volume additive effects can be significant when using volumetric procedures. All procedures should be based on adequate calibrations. Calibrated weights should be used, and volumetric apparatus, especially, will need calibration. The propagation of error will need to be considered, especially when the masses or volumes of analyte/surrogate and matrix are significantly different.

The preparation of evaluation samples by blending is straightforward. Either measured amounts of the components or a dilute mixture of the analyte is added to a measured amount of matrix. When a material, A, is diluted with a second material, D, containing a negligible amount of the analyte of interest, the expression to be used in calculating the composition of the mixture is:

$$\alpha_{A+D} = \frac{\alpha_A W_A}{W_A + W_D}$$

where α_{A+D} = weight percent (or ppm) of constituent α in mixture (A + D)
W_A = weight of sample A in mixture
W_D = weight of sample D in mixture
α_A = weight percent (or ppm) of α in material A

If there is a significant concentration of the analyte in material D, the expression to be used is:

$$\alpha_{A+D} = \frac{\alpha_A W_A + \alpha_D W_D}{W_A + W_D}$$

where α_D = weight percent (or ppm) of constituent α in material D and all other symbols are the same as before.

Liquid Samples

Miscible liquids blend perfectly, but stratification of the several constituents can result from the mode of addition. Thus, bulk additions of one liquid into another could result in layers, especially when they differ significantly in density. In an opaque container, the existence of multilayers could go undetected. Careful mixing should take care of these problems and must be done.

Solid Samples

The spiking of solid samples can take place by either liquid or solid addition. In the former case, a solution of the analyte desired may be added. As the solvent evaporates, a deposit of analyte may be left in a localized area of the matrix. Such samples are difficult to mix uniformly so the entire sample may need to be consumed (see [134]). When solids are mixed, the two materials must be reduced to approximately the same small particle size if they are to be reasonably blended [77]. The uniform mixing of a small amount of powder A with a much larger amount of power B is no trivial exercise. Again, consumption of the entire mixture may be advisable.

Gas Samples

Gases may be blended statically by the mixing of gravimetrically (preferred), volumetrically, or manometrically measured quantities. While gases mix completely, considerable time may be required to overcome stratification caused by introduction of disproportionate quantities of the components.

Dynamic blending of two gas streams, in which flow rates are carefully controlled and measured, is an excellent way to produce gas mixtures. The blending system should be designed to produce turbulence at the point of mixing. In such cases, uniform mixing is achieved at a downstream distance of 7 to 10 diameters of the flow system tubing.

STABILITY OF MIXTURES

The fact that certain components have been quantitatively introduced is no guarantee that they will be present or remain in that proportion in a mixture. Preferential adsorption (or reaction) at the walls of the container may prevent this; hence, choice of the container is a critical decision. Likewise, leaching or dissolution of constituents from the container walls can cause compositional problems. Such considerations often suggest the dynamic dilution process as the best way to prepare gaseous and liquid evaluation samples, but this may not be a viable approach.

Because of the reactivity of some gases and differential reactivity of components of mixtures, one cannot assume that the resulting blends will contain the exact proportions added to the container. Ways to investigate this problem are discussed in a recent paper by Dorko and Hughes[44a].

Interaction of constituents with oxygen and radiation may need to be considered. The necessity for low temperature storage may be another complicating feature. If test samples have any problems in the above matters, the evaluation samples will need the same kind of protection, perhaps to a greater degree.

SUBSAMPLING

Ordinarily, an evaluation sample will be larger than the analytical sample. In such cases, some subsampling will be required. The supplier will need to verify that a representative subsample can be taken; otherwise, or else the entire evaluation sample may need to be consumed. The minimum sample size needed to realize the certified value of analyte ordinarily will need to be specified. In the case of certified reference materials, the supplier should advise the user on such matters. Due consideration must be given to such problems when preparing evaluation samples. If the entire sample must be consumed and is inappropriately large, handling it can involve a different set of problems than for the test sample (e.g., larger blanks), and this can confound the conclusions.

Specific instructions for subsampling will need to be supplied with each evaluation sample. Otherwise, variability in subsampling could cause overriding variance problems.

INTEGRITY OF EVALUATION SAMPLES

Evaluation samples must be beyond reproach. They must be both technically sound and defensible. Any lack of confidence in them can not only cast doubt on present results but also provide the basis for doubt on future samples. When evaluation samples are prepared by a quality assurance laboratory, the laboratory should run continuous control of its production process. It should acknowledge any problems and recall doubtful samples. Legitimate questions should be resolved experimentally.

The reported values for analytes in evaluation samples may be based on preparative data or on analyses. In either case, the data must be technically sound and defensible. Only procedures recognized as being capable of providing reliable results should be used. The procedures used in certifying SRMs, described in Chapter 17, should be used to the extent possible and feasible.

CHAPTER 17

Reference Materials

In the most general terminology, a reference material (RM) is a substance for which one or more properties are established sufficiently well to calibrate a chemical analyzer or to validate a measurement process [34, 70, 135]. An internal reference material (IRM) is such a material developed by a laboratory for its own internal use. An external reference material (ERM) is one provided by someone other than the end user. A certified reference material (CRM) is a RM issued and certified by an organization generally accepted to be technically competent. A Standard Reference Material (SRM) is a certified reference material issued by the National Bureau of Standards (NBS).

ROLE OF REFERENCE MATERIALS

A reference material is used in a decision process; hence, the requirement for reliability of the value of the property measured must be consistent with the risk associated with a wrong decision. The appropriateness of the reference material in the decision process also must be considered. For some purposes, a simple substance, mixture, or solution will be adequate, and the value of the property may be calculated from the data for its preparation. However, even this is best verified by suitable check measurements to avoid blunders. Many decision processes require a natural matrix reference material which may necessitate extensive blending and homogenization treatments and complex analytical measurements. In such cases, only a highly competent organization may have the resources and experience to do the necessary work.

The terms *certificate* and *certification* merely refer to the documentation that supports the reference material. Guidelines for the content of certificates for reference materials have been prepared by the International Standards Organization [71]. They recommend the kind of information the certificate should contain but do not describe how it should be obtained. Furthermore, there are no guidelines for judging the relative quality of reference materials. Accordingly, users of reference materials are urged to seek documented evidence from the producer and to evaluate the reliability of any reference material used in a critical measurement situation.

Table 17.1. Uses of Reference Materials in Measurement Systems

Method Development and Evaluation
Verification and evaluation of precision and accuracy of test methods
Development of reference test methods
Evaluation of field methods
Validation of methods for a specific use
Establishment of Measurement Traceability
Development of secondary reference materials
Development of traceability protocols
Direct field use
Assurance of Measurement Compatibility
Direct calibration of methods and instrumentation
Internal (intralaboratory) quality assurance
External (interlaboratory) quality assurance

USES OF REFERENCE MATERIALS

Reference materials find a wide variety of uses in analytical chemistry, as illustrated in Table 17.1.

Any use of a reference material depends on the ability to make valid inferences from the measurement results. This involves the tacit assumption or demonstrated evidence that the material is reliable and capable of challenging the measurement process. Furthermore, the measurement process must be known to be in a state of statistical control, since limited measurements of the reference material will be used for predictive or evaluative purposes. Because of the inferential nature of the decisions involved, the reference material should simulate actual test samples both with respect to matrix and level of analyte. When there is a significant mismatch in either of these characteristics, the confidence in any conclusions drawn from the measurements may diminish.

Standard Reference Materials (SRMs) and their counterpart CRMs are used widely to assure compatible data. Data that is compatible has attained peer status with respect to accuracy and can be used with other data of the same kind for appropriate decision purposes. The rationale behind the use of reference materials to assure compatibility is illustrated in Figure 17.1. An organization such as NBS develops and certifies accurate values for analytes in an appropriate SRM. Laboratories measure the SRM, observing recognized quality assurance practices.

To the extent that they can produce acceptable measurement results, all laboratories using a specific SRM may be said to be intercalibrated with other laboratories so performing, and indeed with NBS, in so far as measurement of the specific SRM is concerned. To the extent that the SRM is applicable, the

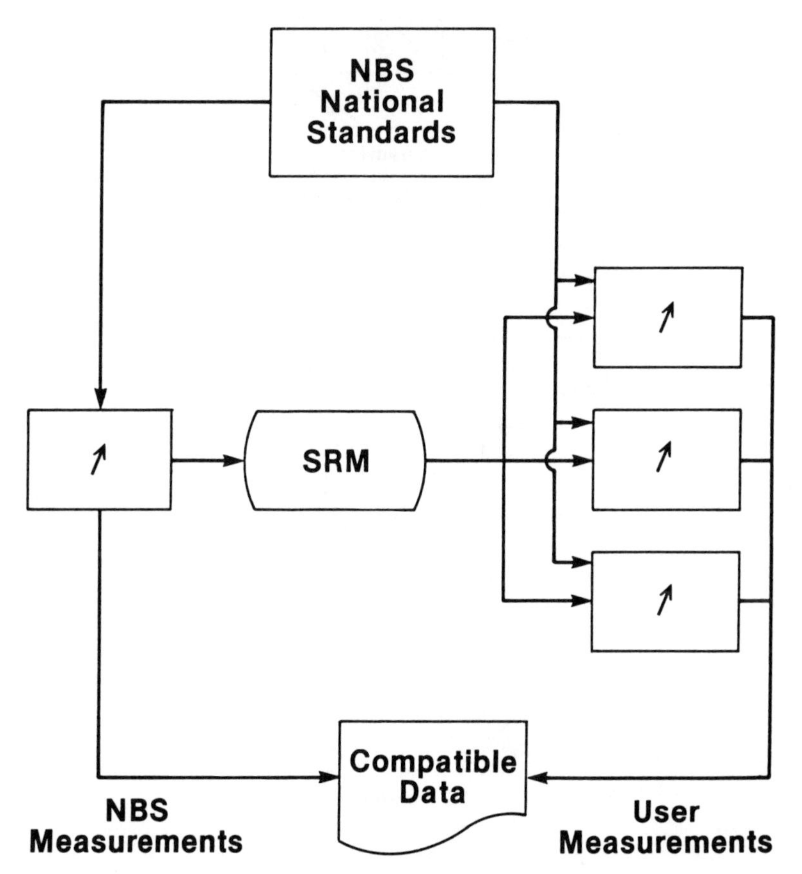

Figure 17.1. Measurement compatibility by intercalibration, using an SRM.

concept of intercalibration and production of compatible data can be transferred to the measurement of other samples.

In considering the validity of any claim for intercalibration and compatibility, the question of the appropriateness of the SRM must be answered satisfactorily. Any laboratory making such claims has the burden of proof for such claims.

The four general cases for use of SRMs as quality assurance materials are illustrated in Figure 17.2 (A-D). When a matrix match is possible (A), the uncertainty in the sample measurements can be equatable to that observed in measurement of the SRM. When such a match is not possible, but an SRM with a related matrix is available (B), the test sample uncertainty may be relatable to those observed when measuring the SRM. Even when the above situations do not apply, the measurement of an appropriate SRM (C) can monitor the measurement system, and its performance when measuring test samples can be inferred in many cases. When an SRM is unavailable or not

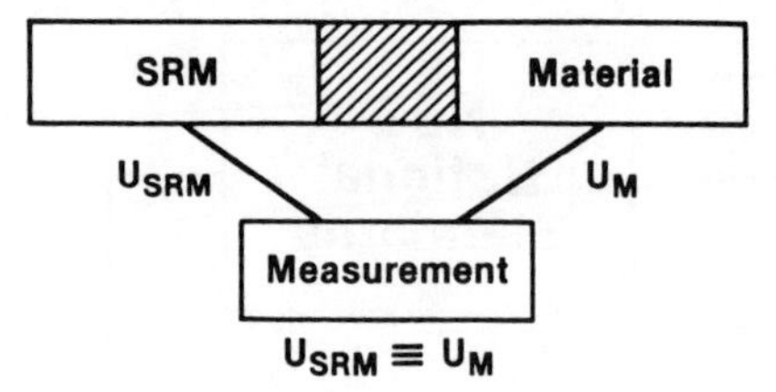

(A) Matrix-Match

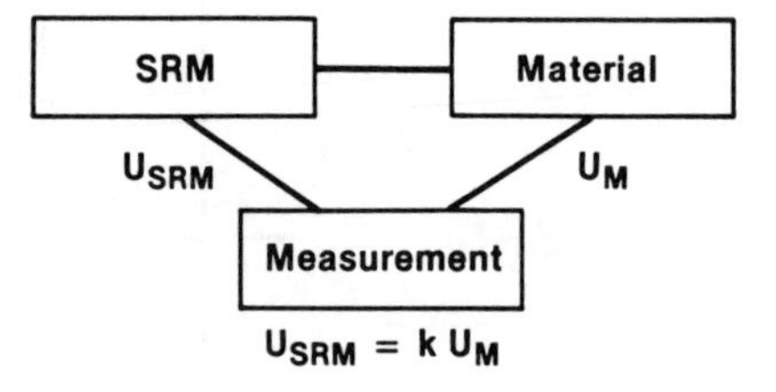

(B) Matrix-Related

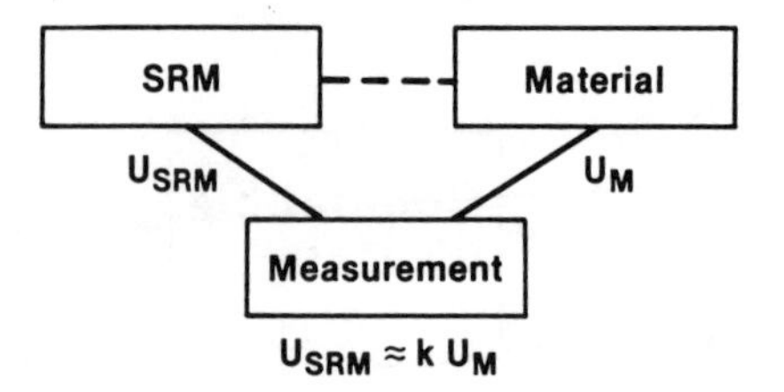

(C) Matrix-Relation Inferred

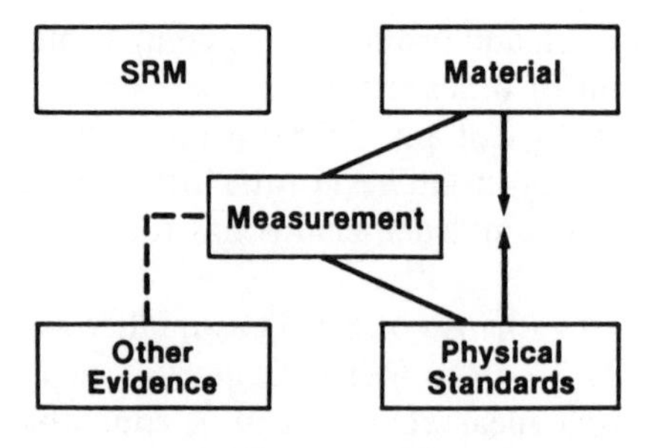

(D) SRM — Not Used

Figure 17.2. Interpreting SRM measurements.

used (D), measurement uncertainty must be inferred from other evidence, such as physical calibrations and the experience of others.

When used to assure measurement compatibility, the analysis of a single reference material may not fully diagnose a measurement system. As discussed in Chapter 24, several reference materials, differing in analyte level, may be needed to assure the absence of significant bias throughout the measurement range or to identify its nature if found.

The continuing use of reliable reference materials utilizing control charts is recommended for all critical measurement situations. The planned collateral measurement of accurately certified (and hence expensive) reference materials and internal reference materials, using collateral control charts, is a good approach to monitoring a measurement process.

CERTIFICATION OF REFERENCE MATERIALS

The most useful reference materials are those for which some property(s) is certified on the basis of its accuracy. This is because the chief role of an RM is to evaluate accuracy, since precision can be easily estimated by other approaches. A certified value of known accuracy can be obtained in several ways, as discussed below.

Whenever possible, a definitive method should be used (see Chapter 9). In order to minimize the chance of misuse, measurements should be made independently by at least two different analysts. There should be no statistically significant differences in the results. When this approach is infeasible or impossible, the analytes of concern should be analyzed by at least two independent measurement techniques, again without significant differences. Limits of uncertainty should be assigned to the certified values, according to the principles described in Chapter 22. Needless to say, the experimental work in each case should be well planned following appropriate quality assurance practices.

Because material variability must be evaluated, a statistically based sampling plan should be followed. Control charts should be maintained during the course of the measurements to ascertain statistical control.

The stability of reference materials is as important a consideration as are their certified values. Stability studies should precede any certification measurements if there are any doubts on this matter.

In the case of solid samples, attention should be given to the minimum size of sample that can be expected to provide results consistent with the certified values. If there are any precautions that need to be taken with respect to drying, mixing, or other preliminary operations, these must be defined.

Additional information on the preparation and use of reference materials will be found in the *Handbook for SRM Users* [134].

CHAPTER 18

Traceability of Chemical Measurements

The terms "traceability" and "traceable to national standards maintained by NBS" and even "traceable to NBS" are receiving increasing usage. The general concept of traceability has been discussed by Belanger [22] with special emphasis on calibrations and physical measurements. The present discussion concerns its application to chemical measurements.

Traceability literally means the ability to trace, and hence implies the existence of an unbroken, identifiable, and demonstrable pathway. Moreover, the measurement process must be fully documented, and all sources of error identified and quantitatively evaluated. In other words, traceability can be likened to the genealogy of a measurement process. It is a prerequisite for the assignment of limits of uncertainty but does not imply any level of quality. Thus, two similar standards can both be traceable but differ widely in their quality.

The term "standard" is used throughout this discussion in a generic sense. It can denote anything from fundamental standards used to define the basic units of measurement, to a material, substance, instrument, or scale used to establish a numerical value in the most mundane measurement system. No matter how it is used, the standard must have sufficient accuracy within the requirements for its use.

CHEMICAL TRACEABILITY

The purpose of all claims for traceability is to establish the accuracy of measurements. Measurement almost always consists of comparison of an unknown, the value of which is desired, with a standard, the value of which is believed to be known with a requisite degree of reliability. In physical measurements, the known is often an object or a scale that has been calibrated against a standard object or scale which could have been several generations removed from national or primary standards. The uncertainty in the standard or scale used can be known only if the history of all intermediate calibration measurements is known. This includes the specific standards that were used and the uncertainties of all intercomparisons. In essence, traceability is the capability of reconstructing the chain of events and the assignment of a final statistically supportable total measurement uncertainty to any standard that is used in a measurement process. If any link is missing (e.g., the standards used or the corresponding measurement data), measurement uncertainties cannot be assigned, and the measured value is not traceable.

While chemical measurements are usually more complex than physical measurements, the concept of traceability is essentially similar. Chemical properties are measured with reference to those of known chemical standards. This may be a direct comparison but is usually accomplished by means of scales or instruments that have been calibrated with respect to knowns. Herein lies an often overlooked component of uncertainty of chemical measurements. The standard may fulfill all of the requirements for traceability but may not correspond fully to the unknowns, and full correspondence may be very difficult (if not impossible) to achieve in a given situation. Accordingly, the analyst must use methodology with minimum matrix effects, use matrix modifiers, apply corrections for matrix effects, or remove the component of interest from the matrix (by extraction, dissolution, distillation, etc.) before measurement. All of the above steps introduce uncertainties into the measurement process that must be evaluated, but the evaluation may involve considerable uncertainties as well.

The uncertainty of the chemical standards used may be considered from two points of view: appropriateness and accuracy. The first of these is a qualitative judgment and precedes all other requirements. In many cases, the standard used only calibrates the final measurement step; intermediate steps are not calibrated, yet they can introduce major sources of measurement uncertainty.

Uncertainties of the compositional accuracy of the standards used may be a critical factor in some cases and not in others, but this must always be considered. Here is where traceability enters the picture. When standards are made by anyone other than the analyst using them, they should not be taken for granted. Only standards with documented uncertainties should be used; this should be the policy of every laboratory. It is the obligation of every supplier of standards to provide the information described above.

In many measurement situations, the major uncertainty is not in the standards used but in the performance of the systems. Because of its complexity, a measurement system may have many sources of error and may need to be stabilized by a strict system of quality control. It cannot be overemphasized that statistical control must be attained and maintained if the measurements are to have any logical significance (see Chapter 3). The efficacy of the control procedures and overall quality of the data need to be monitored and evaluated by quality assessment techniques. Reference materials play key roles in quality assessment in such activities as control samples (using control charts), proficiency tests, and performance audits.

Questions of appropriateness of a reference material and accuracy of its compositional values are again of prime importance. Thus, the concept of traceability applies to reference materials as it does to standards for measurement in general.

Standards Prepared by the Analyst

The analyst who prepares standards for her/his own use has the responsibility for their accuracy and for all traceability considerations involved. Obviously, only chemical constituents of requisite and documented purity should be used. Such documentation becomes part of the traceability chain. Moreover, all physical measurements involved, including mass, volume, pressure, and temperature, for example, need to be traceable. Equally important, the containers used and their cleanliness need to be considered as well as the stability of the standards, after they have been prepared. If there are gaps in any of the above information base, the standards may lose their traceability and even their integrity.

Standards Prepared by Others

The largest area of concern for traceability is for standards produced by someone other than the user. More and more, calibration standards are being provided by vendors, and reference materials are almost always so obtained. The vendor of standards and reference materials has the responsibility for the traceability of all certified values. The various aspects of traceability as related to purveyors of such materials are discussed in the following sections.

PHYSICAL STANDARDS

The physical standards such as volumetric apparatus and masses, commonly used in preparing chemical standards, are available from vendors. Ordinarily, such standards can be obtained with certified values that are traceable to NBS, or they can be calibrated to provide such traceability. NBS provides certain calibration services [75] or can advise where such are available. State Weights and Measures Laboratories provide a number of calibration services which are traceable, and ASTM standard methods are available for many of the calibrations that the analyst may need. In all of the above cases, the burden of proof for traceability lies with the person or organization providing the calibration service.

REFERENCE MATERIALS PRODUCED BY NBS

All reference materials supplied by NBS follow the traceability principles discussed in this chapter[134]. NBS provides many SRMs that can be used as chemical standards. They consist of pure chemicals, buffers, and compositional standards that either are standards in themselves or serve as the starting materials for the preparation of chemical standards by the user. The NBS certificate lists the certified values and their uncertainties. When used by others to prepare standards, additional uncertainties may be introduced that must be evaluated by the preparer.

The largest fraction of SRMs are used as quality assurance materials. That is, they are measured by the user to evaluate some aspect of the measurement process. The NBS certificate states the uncertainty of the certified value including that due to its measurement and that due to any between-sample variance (heterogeneity). In some cases, stability limitations are defined, and the user may need to exert certain precautions such as mixing, drying, and storage before the limits of uncertainty are realizable. Of course, the user has the full responsibility for observing these precautions and for providing evidence of their observance in the case of any question of traceability. Also, the user has the responsibility of deciding whether the SRM is appropriate for its use and for defending any inferences made from the measurement data. The user must defend the quality assurance aspects of any measurements made of an SRM when claims for traceability are made.

REFERENCE MATERIALS PRODUCED BY OTHERS

The uncertainty of any reference material or chemical standard provided by any vendor should be furnished to the user. The principles for estimation of the uncertainties are the same, no matter who is the producer. Accordingly, it is recommended that all vendors follow the general guidelines described in this chapter and in Chapter 22 when reporting uncertainties and their traceability. When an NBS/SRM is one link in the traceability chain, the limitations described elsewhere in this chapter apply as well.

The producer of reference materials has the responsibility for the appropriateness of the SRM used for either standardization or quality assurance of the certification process. In addition, it has the responsibility for statistical control of the measurement process and for the statistical accuracy of the uncertainties claimed for the certified values. Questions of stability and homogeneity must also be addressed when certifying values for the material supplied.

A class of reference materials called "Certified Reference Materials" (CRMs) is being produced commercially. These CRMs have a close resemblance to and are certified with respect to SRMs. Because of their importance as complements to SRMs, a protocol has been developed [63] by which reliable certified values and traceability can be established. While the vendor has the final responsibility for measurements involved, the protocol, when faithfully followed, defends the type of evidence that is deemed to be sufficient to support claims for traceability.

TRACEABILITY OF ANALYTICAL RESULTS

The concept of traceability can be extended to every analytical measurement, whether made to certify the composition of standards or for the most practical purpose. The estimated accuracy of any analytical result must be known, and this must be within the requirements of the application; otherwise, the data cannot be used. The measurement laboratory has the responsibility for the accuracy of its measurements, which is best supported by reliable calibrants and quality assurance practices. Again, the reliability and appropriateness of the calibration procedures, and reference materials and the statistical control for the measurement practices used are the responsibilities of the analyst. When an NBS/SRM is used in the measurement process, only the certified value of the SRM is traceable to NBS. All other aspects of the traceability chain used to support claims of accuracy of the data are the responsibility of the analyst.

All aspects of work done within a laboratory must have internal traceability. This means that every detail of the measurement process must be documented and easily retrievable. Data must be traceable to the actual samples measured, the specific instrumentation used, and the specific operator. Calibration standards must be traceable to their source and method of preparation. Calculated results must be traceable to the raw data. The source of all supplemental data must be identified.

CONCEPT OF THE NATIONAL MEASUREMENT SYSTEM

To be compatible, all measurements must be made with respect to a logical and consistent system of units maintained by appropriate standards. Metrologists have recognized this need for many years, and international conventions and treaties have developed, Le Systeme International d'Unites (SI) [99A], which establishes seven fundamental units upon which all others may be based. Moreover, standards have been agreed upon to maintain these units. The National Bureau of Standards maintains measurement standards for the United States that are in harmony with those of all major countries. Virtually all physical measurements can be made using SI or metric units maintained by standards that are traceable to NBS. A full discussion of this subject is beyond the scope of the present book, but it is presented excellently by McNish [93].

Measurement compatibility is enhanced as all use the same units of measurement. The use of the metric system, and more precisely, the SI system of measurement units is recommended. In any case, the units used to express values must be identified and stated explicitly. When this is done, and when compatible and traceable standards are used for all calibrations, measurements can be interrelated, no matter what system of units is used. The compatibility of data further requires that measurement uncertainty be evaluated for all numerical data.

The national measurement system for analytical chemistry is shown schematically in Figure 18.1. Chemical measurements directly or indirectly involve physical measurements as well. The physical basis for chemical measurements is illustrated in the upper portion of the figure. Standards for the calibration of all necessary physical measurements are available, and the measurements

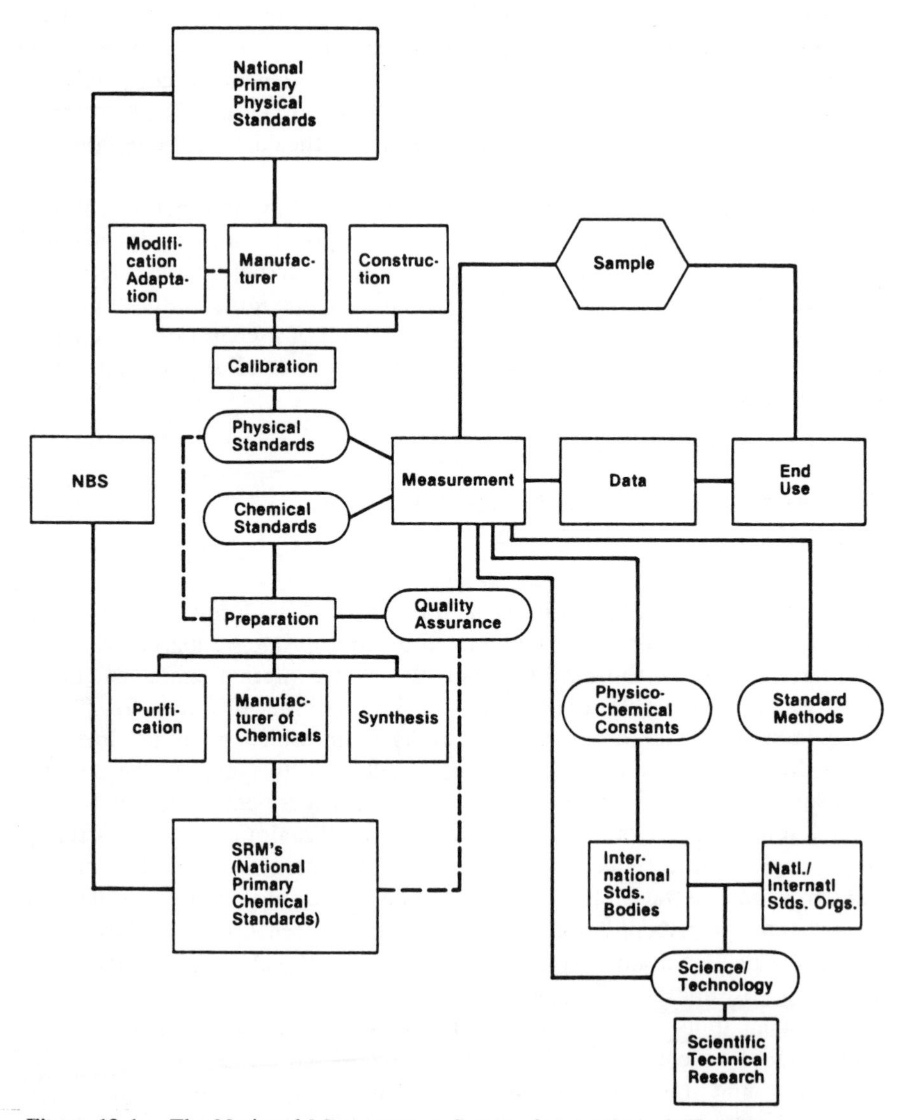

Figure 18.1. The National Measurement System for Analytical Chemistry.

themselves can be made within the accuracy requirements of chemical data, provided traceability and reasonable quality assurance are practiced.

There is no system for providing chemical standards for calibration of chemical measurements that parallels the situation for physical measurements. Due to the wide variety of measurements that are made, it would be practically impossible to do so. Chemical standards necessary for chemical analysis traditionally have been prepared by the analyst, but this situation is undergoing rapid change. More and more, suppliers of standards or the materials used to prepare them are entering the picture. Such suppliers have a profound responsibility to provide reliable wares and full support for their traceability. No supplier can, in good conscience, market low-quality standards, and no laboratory should use any standards except those demonstrated to be adequate for the intended use.

The question of suitability of the standards used, discussed in Chapter 10, is one that must be answered by every analyst. Standard Reference Materials, as available, may be considered as part of the national measurement system. Their role, discussed more fully in Chapter 17, is to assist in the evaluation of the accuracy of the measurement process and in the validation of methodology.

CHAPTER 19

Quality Audits

Quality audits are considered an essential part of a quality assurance program. Audits are of two types: systems audits describe those conducted to qualitatively evaluate the operational details of a QA program, while performance audits are conducted to quantitatively evaluate the outputs of a measurement system. Both may be internal or external, denoting whether the activity is conducted by the laboratory itself or by someone not connected with it.

A systems audit should be conducted with respect to some type of expectation. This may be based on generic criteria such as those developed by ASTM [8, 10, 11, 17] or the American Council of Independent Laboratories [1]. The degree of compliance with a laboratory's own quality assurance program may be another objective. In some cases, an audit may be conducted to evaluate the degree of compliance with mandatory requirements such as those necessary to attain or maintain accreditation or a license.

Performance audits evaluate the quality of data outputs with respect to the state of the art or with legally imposed limits, or for maintenance of a laboratory's own performance standards.

INTERNAL AUDITS

Internal Systems Audits

The simplest type of internal system audit consists of an ongoing supervisory surveillance of the quality assurance practices of subordinates. Periodic notebook checks are an example of a routine audit of procedural details. Supervisors are ordinarily responsible for assuring that GLPs, GMPs, and SOPs are followed, and a systematic procedure to ascertain this may be considered an internal audit. At the same time, it should be ascertained that all required records are kept in a satisfactory and readily understandable fashion.

Audits may be conducted periodically or aperiodically by the quality assurance officer and/or by an auditing committee to monitor quality assurance practices. Such audits may be either complete or random [53]. Overemphasis on the auditing of perceived trouble spots could result in overlooking problems in preconceived trouble-free areas. When spots for audits are randomly selected, the latter situation is avoided. Also, the staff is less likely to consider an audit as a witch-hunt and hence be more receptive to the idea.

Audits can consist of checks of critical operations in randomly selected areas. When testing is a major activity, randomly selected test records can be reviewed to see that all required documentation is available. Some tests should be examined in detail to determine the quality as well as the completeness of the information.

The frequency and criticality of internal audits will depend on the critical nature of the data outputs, on the threat or risk associated with erroneous or questionable results, and on the pride of workmanship of management and staff. A quarterly schedule is recommended initially, with adjustment to a semiannual basis as justified. Of course, supervisory audits are ongoing and more frequent.

Whenever a laboratory is subjected to external audits, its internal audits are advisedly more frequent and more critical than the former. There should be no surprises resulting from external audits. In some laboratories, there are mandatory operations that must be done on a scheduled basis. Examples include critical calibration intervals and expiration dates of specific reagents. A computer program, a checklist, or other systematic approaches may be used to warn of impending deadlines and to check that they have been observed. While a minor slip of such a deadline may not seriously affect results in a technical manner, it could cause problems of defensibility.

Internal Performance Audits

An internal performance audit consists essentially of reviewing the ongoing quality assessment program of a laboratory. Its objective should be to evaluate the accuracy of all data. Control charts may be reviewed to assure that they are up-to-date and that control samples are being measured at specified intervals. The status of reference materials can also be checked. Some large laboratories have a quality assurance group that prepares spikes, split samples, and other kinds of control samples and statistically analyzes the results of their measurement. There needs to be a high degree of cooperation between the bench and such a group if the assessment is to be realistic. Needless to say, a laboratory is responsible for the quality of its data and must take reasonable efforts to assure itself and interested parties of the confidence that can be placed in it. When this is done, it should pass any external performance test with flying colors.

Mechanism

Internal audits may be conducted by a quality assurance officer, a team, or a combination of these. Audits should be as objective as possible; hence, the use of an interdisciplinary audit team has distinct advantages. Because service on an auditing team can be instructive, the membership of a group should be rotated. A group member may abstain whenever her/his activity is audited. The interdisciplinary approach provides a good basis for identifying both desirable parallelisms and justifiable differences in the quality assurance practices within a laboratory.

Corrective Actions

The purpose of any audit is to verify both compliance and performance, and to identify discrepancies when they exist. In the latter case, any problems should be addressed and remedied in an appropriate manner. Problems can ordinarily be classified as undesirable but not critical or as critical and requiring immediate action. The remedy could be closer adherence to some aspect(s) of the quality assurance program, amendment of the program, or incorporation of features not already addressed.

The maintenance of a log of problems encountered and the corrective actions taken is essential if an auditing program is to be successful. This log should be reviewed for recurrence of the same problems and for the presence of patterns that could be symptoms of broader problems. Hopefully, periodic reviews will show the elimination of any serious problems and the decrease of problems overall.

The prompt addressing of major and even minor problems identified by an audit is necessary. There should be no carry over from audit to audit. Punitive audits should be discouraged, or else they precipitate cover-up actions. When quality circles exist, they may be used as the means to identify the corrective actions that would be most beneficial.

EXTERNAL AUDITS

External quality audits should follow essentially the format described for internal audits. Ordinarily, they will be conducted to ascertain compliance with rules, regulations, or criteria for certification, and will have a higher degree of formality than the former. Where mandatory records are required, compliance with such will be critically evaluated. The search for any corrective actions and the correction of problems identified in a previous audit will be an important activity.

External auditors ordinarily want to see the records of internal audits, especially if they are required in a mandatory quality assurance program. Well-maintained records can inspire the confidence of external auditors as well as facilitate and increase the effectiveness of external audits. Auditors also want to review the control charts maintained by a laboratory. The ease with which important records and information can be retrieved is a criterion for judgement of the management practices of a laboratory.

INDEPENDENT AUDITS

Periodic audits or surveys by an independent expert can be beneficial to management in its ongoing implementation of its quality assurance program. Such an audit or survey can also be helpful when a laboratory is developing a QA program or revising an existing one. Scheduled independent audits can be used as controls on internal auditing programs.

An independent audit is a confidential service to management and should provide opinions or recommendations that management may use within its own discretion. An independent auditor should perform such services with high ethical standards as befit any consultant to management.

FORMS

This section contains forms that have been found useful when evaluating QA programs. They may be used in several ways, some of which are suggested in the following sections.

Personal Quality Assurance Profile

This checklist is designed for individuals in a self-appraisal process. When an appraiser is provided with copies of such, this information must be treated with the highest confidentiality.

Laboratory Quality Assurance Profile

This checklist is designed for supervisors and senior staff members to evaluate the level of existing QA activities. When used in connection with the first form, it can compare the perceptions of management and staff on matters of quality assurance. Copies of completed forms may be given to an independent auditor to assist in the appraisal of existing activities.

Quality Assurance Survey

This is an outline that nay be followed by anyone conducting a survey of the QA practices of a laboratory.

Quality Assurance Practices Checklist

This checklist may be found useful when conducting a survey or audit.

QA Training Profile

This checklist is useful for identifying the extent of QA knowledge of individual staff members. The staff may use it as a means of self-appraisal, or management may use it in deciding what specific training to provide to its staff.

Form No. 1

PERSONAL QUALITY ASSURANCE PROFILE
Self-Appraisal

This checklist is designed for use of individuals to check their own QA profile by the self-appraisal process. Each item should be considered, and the score for the statement best describing the expertise/knowledge/performance should be entered in the box. Intermediate values may be chosen as appropriate. The level determinants are meant to be suggestive and are open to interpretation.

Tabulate the average score as indicated. An average score of 3.8 is acceptable but not laudatory. A score of 2.5 or lower is considered to be unacceptable and indicates major QA deficiencies. Intermediate scores indicate the need for immediate remedial actions.

No matter what average score is obtained, individuals should examine the scores for individual items to identify QA areas that need improvement. Any item rated at 3 or below should be so considered.

[ ] I. Knowledge of Field

State-of-the-art knowledge of my field	5
Good practical knowledge of field	3
Some gaps/limited understanding	1

[ ] 2. General Understanding of Methodology Used

Excellent comprehensive knowledge of methodology, including basic theory	5
Make point to understand methodology before use	3
Practical understanding but some gaps in basic comprehension	1

[ ] 3. Mastery of Specific Technology

State-of-the-art accuracy and precision always attained	5
Average accuracy and precision attained	3
Accuracy and precision needs improvement	1

[ ] 4. Use of Written SOPs

Use SOPs and/or develop written procedures for all methods used	5
Use SOPs/written methods for all critical analyses	3
Limited or little use of written methods	1

[ ] 5. Precheck of Methodology Prior to Use

Extensive checks of new methods/preanalysis check of all others before use	5
Prechecks confined to new methodology	3
Little or no prechecking	1

[ ] 6. Adherence to GLPs/GMPs

GLPs/GMPs developed and used regularly	5
GLPs/GMPs for most critical operations	3

GLPs/GMPs little/not used 1

[ ] 7. Laboratory Notebooks
- Neat, indexed lab notebooks maintained, readily understandable to others 5
- Some deficiencies in notebooks but generally acceptable 3
- Notebooks need considerable improvement 1

[ ] 8. Knowledge of Statistics
- Full working knowledge and extensive use of statistics in all decision processes 5
- Good understanding and occasional use of statistics 3
- Limited knowledge of statistics 1

[ ] 9. Control Chart Usage
- Good understanding, extensive use of control charts 5
- Occasional use of control charts 3
- Limited/little use of control charts 1

[ ] 10. Participation in Technical Activities
- Active/leadership role in technical organizations 5
- Passive role 3
- Little participation 1

[ ] 11. Training Courses
- Training course(s) taken during past 18 months 5
- Training course(s) taken during past 3 years 3
- No recent training taken 1

[ ] 12. Technical Books/Informal Training
- Informal training/tech books read during past year 5
- Some informal training during past 2 years 3
- No recent informal training 1

[ ] 13. Experimental Planning
- Experimental planning understood/used extensively in all major activities 5
- Work generally well-planned before starting 3
- Planning needs considerable improvement 1

[ ] 14. Use of Randomization
- Understand/always use randomization in work plans/execution of work 5
- Some use of randomization in work plans 3
- Limited/little use of randomization concepts 1

[ ] 15. Housekeeping Practices
- Work space always tidy consistent with activities in progress 5
- No major problems in housekeeping 3
- Housekeeping not a strong point 1

Average Score [ ]

Form No. 2

LABORATORY QUALITY ASSURANCE PROFILE
Self Appraisal

This checklist is designed for use by laboratory management to appraise the program of the laboratory. Each item should be considered individually and the appropriate score entered in the box. Intermediate values may be chosen. The level determinants are meant to be suggestive and are open to interpretation.

An average score of 3.8 is acceptable but not laudatory. A score of 2.5 or lower is considered to be unacceptable and indicates that serious risk exists in laboratory operations. Intermediate scores will require a review of the QA program to identify and rectify major deficiencies.

Even in the case of an acceptable average score, a low score for any item (<3) should be considered for possible corrective actions.

[ ] 1. Laboratory QA Program

Written plan adopted/implemented/in use	5
Definite but informal program	3
Informal/variable program	1

[ ] 2. Use of Written (Before Use) Methodology

Exclusively	5
Majority of time/for all critical data	3
Few or none used	1

[ ] 3. Control Chart Use

Maintained for all critical operations	5
Variable but significant use in organization	3
Little or no use	1

[ ] 4. Uncertainty Limits for Data

Limits for all data outputs/policy enforced	5
Most of the time/at least where critical	3
Minority of cases	1

[ ] 5. Reports/Proposals

Pre- and post-screened for QA aspects	5
Those deemed critical are screened	3
Variable/seldom done	1

[ ] 6. Facilities Maintenance

Excellent (showplace condition)	5
Good (passes muster)	3
Poor (reservations, no-no areas)	1

[ ] 7. Equipment Maintenance

Regular maintenance with records kept, control charts as appropriate	5
Good maintenance, documentation of such has some deficiencies	3
Irregular maintenance practices	1

[ ] 8. Records

Laboratory records judged excellent by any standards	5
Some reservations, could be difficulties in spots	3
Variable, need considerable improvement	1

[ ] 9. Training New Employees

Formal QA indoctrination	5
Informal QA indoctrination	3
Assumed not needed/not done	1

[ ] 10. Personnel/Staff QA Consciousness*

Staff average for personnel QA audit >4	5
Staff average for personnel QA audit 3 to 4	3
Staff average for personnel QA audit <2.5	1

*Alternatively, average staff QA audit values may be inserted in

[ ] 11. Professional Interactions

Majority and all key staff active in some professional organization	5
Reasonable level of activity	3
Little or don't know	1

[ ] 12. Management and Statistics

High level of knowledge and ability to use at supervisory level and above	5
General awareness and reasonable usage	3
Variable comprehension/use	1

[ ] 13. Internal QA Audits

Regular program/feedback/corrective actions	5
Occasional audits	3
Few or none	1

[ ] 14. External QA Audits

Regular external appraisal of QA policy/practices	5
QA appraisal definite part of other reviews	3
None	1

[ ] 15. Overall Opinion of QA Status

No known weaknesses that are not subject of corrective actions	5
Known QA weaknesses but less than vigorous action to correct them	3
Little or no basis for judgement	1

Average Score [ ]

Form No. 3

QUALITY ASSURANCE SURVEY

1. Initial discussion with Management
   - 1.1 To obtain profile of organization: what it does; who it serves; criticality of its outputs; how it is managed; identification of managerial chain; management's overall opinion of the quality of its work
   - 1.2 To identify concerns: perceived problems; perceived strengths; sensitivities; external pressures
   - 1.3 To identify expectations: reason for desiring a survey; expected use of outcome
   - 1.4 To decide on announcement process to staff
2. Pre-site visit activities
   - 2.1 Management appoints local liaison
   - 2.2 Circulation of QA Profile questionnaire to management and to staff
   - 2.3 Analysis of returned questionnaires (if used) and intercomparison of management/staff opinions
3. Site visit
   - 3.1 Discussion with management of findings to date
   - 3.2 Discussion with management of proposed survey procedure
   - 3.3 Concurrence or revision of survey procedure
   - 3.4 Random selection of survey locales
   - 3.5 Comparison of practices with profile
   - 3.6 In-depth review of some locales
   - 3.7 Identification of strengths and weakness
   - 3.8 Exit discussion with management
4. Preparation of Report of Findings
   - 4.1 Summary of findings
   - 4.2 General QA health of organization
   - 4.3 Present practices that are endorsed
   - 4.4 Good practices needing strengthening
   - 4.5 Minor problems with long-term implications
   - 4.6 Major problems requiring remedial action
   - 4.7 Overall recommendations: Procedural; training; other

Form No. 4

QUALITY ASSURANCE PRACTICES CHECKLIST

1. Administrative Information
   Name of Laboratory
   Organizational Chart
   Scope of Capabilities
   Accreditations, recognitions, organizational memberships

		Yes	No	Exp.
2. Human Resources				
2.1	Roster of personnel available	____	____	____
2.2	Resumes of key personnel available	____	____	____
2.3	Qualifications of all key persons commensurate with responsibilities	____	____	____
2.4	Work load commensurate with staffing	____	____	____
2.5	Staff morale high	____	____	____
3. Physical Resources				
3.1	Facilities adequate for all services offered	____	____	____
3.2	Laboratory environment adequate	____	____	____
3.3	Adequate space for all operations	____	____	____
3.4	Adequate housekeeping	____	____	____
3.5	Adequate safety/inspections	____	____	____
4. Equipment				
4.1	Equipment list maintained	____	____	____
4.2	Equipment used in specific tests identifiable	____	____	____
4.3	Equipment manuals filed systematically: Where ____________________	____	____	____
4.4	All equipment serviceable and in calibration	____	____	____
4.5	Calibration status of all equipment indicated: How ____________________	____	____	____
4.6	List special facilities and opinion of their adequacy ____________________ ____________________			
5. Quality Assurance System				
5.1	QA policy statement on file	____	____	____
5.2	QA Officer with assigned duties	____	____	____
5.3	All employees aware of their QA responsibilities	____	____	____
5.4	QA Manual on file Adequacy ____________________	____	____	____
5.5	Written procedures used for each test	____	____	____
5.6	SOPs for all recurring tests (Methods Manual)	____	____	____
5.7	Staff competent in test procedures	____	____	____

5.8 Methods validated before use: ____ ____ ____
How ____________________

5.9 System to qualify test operators ____ ____ ____

5.10 Reagent control practiced ____ ____ ____

5.11 Demonstration of statistical control: ____ ____ ____
How ____________________

5.12 GLPs, GMPs on file and followed ____ ____ ____

5.13 System of corrective actions ____ ____ ____

6. Control Charts

6.1 QA policy requires use of control charts ____ ____ ____

6.2 Control charts used to monitor system ____ ____ ____

6.3 Control charts used to assign confidence limits ____ ____ ____

6.4 List of current control charts on file: ____ ____ ____
Number on list ____________________

6.5 Control charts available for inspection by clients ____ ____ ____

7. Reference Materials

7.1 QA policy requires use of RMs ____ ____ ____

7.2 Internal reference materials listed and used ____ ____ ____

7.3 All reference materials inventoried ____ ____ ____

7.4 Reference materials used with control charts and limits ____ ____ ____

8. Records

8.1 Systematic format(s) for data ____ ____ ____

8.2 Lab notebooks kept systematically ____ ____ ____

8.3 Charts/readouts referenced to notebooks ____ ____ ____

8.4 Records periodically reviewed for adequacy ____ ____ ____

8.5 All records will pass critical inspection ____ ____ ____

9. Reports

9.1 Reports reviewed before release ____ ____ ____

9.2 Reports reference all important supporting data ____ ____ ____

9.3 Statistically supported limits of uncertainty ____ ____ ____

9.4 System for reports/records/data retrieval ____ ____ ____

10. Sample Management

10.1 Sample management system used ____ ____ ____

10.2 Computerized sample management ____ ____ ____

10.3 Sample preparation facilities adequate ____ ____ ____

10.4 Sample storage adequate ____ ____ ____

11. Audits

11.1 QA system audits on regular basis: ____ ____ ____
Frequency ____________________

11.2 External QA system audits ____ ____ ____

11.3 Performance audits on regular basis ____ ____ ____

11.4 Corrective actions taken as result of audits ____ ____ ____

11.5 Records kept of corrective actions ____ ____ ____

12. Training

12.1 Competence of staff maintained by training ____ ____ ____

12.2 QA training for staff and new employees ____ ____ ____

12.3 QA manual available to all employees ____ ____ ____

12.4 Key staff active in professional/technical/ standardization activities ____ ____ ____

13. Other related information: ______________________________

____

14. Explanation/comments on evaluation above (by item number). Attach additional sheets as necessary.

Evaluation by:

Name ______________________________

Title ______________________________

Date ______________________________

Form No. 5

QUALITY ASSURANCE TRAINING PROFILE
Self-Appraisal

[ ]	1. General knowledge of analytical chemistry	
	Broad/comprehensive	4
	Good comprehension	3
	Average	2
	Deficient	1
[ ]	2. Knowledge of field of activity	
	Authoritative	4
	Above average	3
	Average	2
	Deficient	1
[ ]	3. Knowledge of statistics	
	Facile	4
	Working knowledge	3
	Basic understanding	2
	Deficient	1
[ ]	4. Knowledge/experience with control charts	
	Knowledge and practical experience	4
	General knowledge	3
	Limited knowledge	2
	Deficient	1
[ ]	5. Knowledge/experience in standardization activities	
	Position of leadership	4
	Active participation	3
	Low level of participation	2
	None	1
[ ]	6. Knowledge/experience in general aspects of quality assurance	
	High	4
	Average	3
	Limited	2
	Deficient	1
[ ]	7. Knowledge/experience in sampling	
	Can design sampling plans	4
	Working knowledge	3
	Limited	2
	Deficient	1
[ ]	8. Knowledge/experience in calibration	
	Above average	4
	Average	3
	Limited	2
	Deficient	1

[ ] 9. Knowledge/experience in data evaluation

Can evaluate complex data	4
Can evaluate simple data	3
Limited ability	2
Deficient	1

[ ] 10. Reporting/report writing

Highly skilled	4
Good skills	3
Limited skills	2
Deficient	1

Total Score

Instructions:

Enter in the the numerical score that is believed to describe your capability in each subject area. Use intermediate numbers as appropriate.

Calculate and enter total score where indicated.

Total Score		Subject Area
34–40	Outstanding	4
27–33	Above average	
23–26	Average	2.5
17–22	Below average	
10–16	Deficient	1

If the total score is less than average, the need for general and/or specific quality assurance training may be indicated. When a low score is noted in any subject area, training may be needed if the area is critical to the personal work program.

CHAPTER 20

Quality Circles

A quality circle consists of a small group (ten to twelve maximum) of employees with similar interests or involvements in an activity or process which may benefit from quality improvements. Since almost any activity or process should be concerned with the quality of its outputs, there is virtually no situation for which a circle is not applicable. The group should be reasonably homogenous but not identical in its interest or involvement to provide a sound basis for attack of problems [4, 44].

ORGANIZATION

An organization planning for several quality circles usually will appoint a facilitator who will coordinate and direct circle activities. Each circle will have a leader, usually appointed, but possibly elected by the group. The leader is responsible for the smooth operation of the circle and must involve the participation of all members. Participation of the quiet members is encouraged by asking questions, seeking opinions, and other approaches. Overparticipation of exuberant members is discouraged by the idea-writing approach which will be discussed later.

OBJECTIVES

The objective of a circle is to both prevent and solve problems related to the quality of outputs. While outputs in which the members are directly involved are of major concern, outputs of others related to theirs can also be considered. There is also the possibility of collaboration with others on such problems. The objective is to reduce errors and to enhance quality. Participation encourages cooperation and motivation. Encouragement of ownership of any necessary changes and grass roots inputs to ensure the credibility of quality assurance programs are additional objectives. Circles can become an organizational resource, consisting of teams experienced in troubleshooting. But more than this, they are an excellent way to train members, especially in attitudes

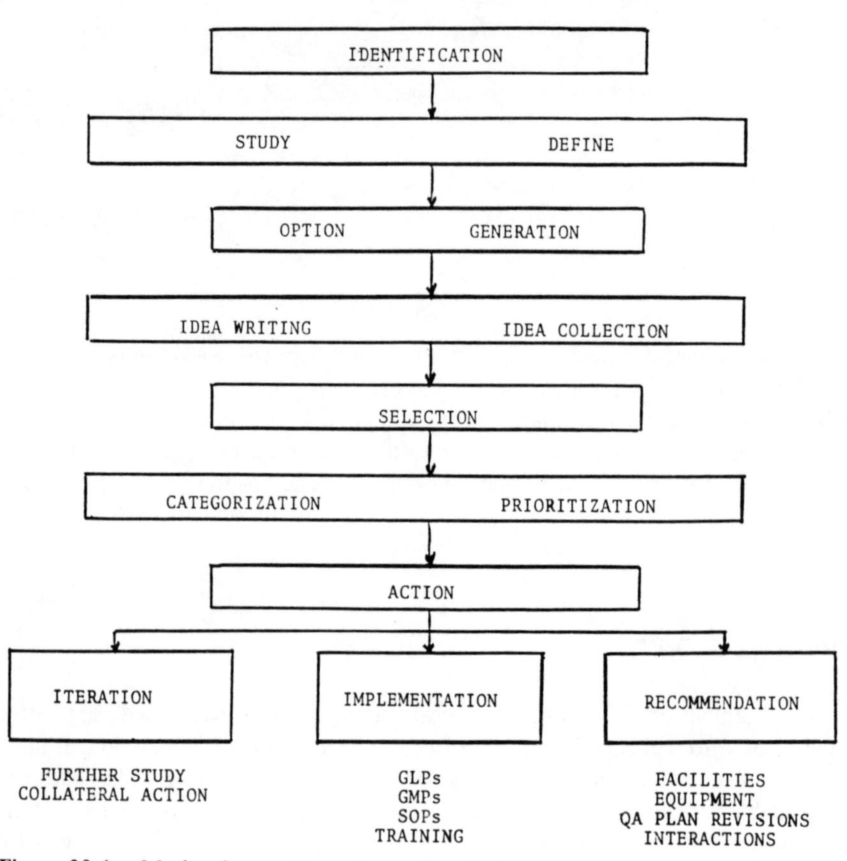

Figure 20.1. Mode of operation of a quality circle.

concerning the quality of outputs. Circles, by virtue of an intensive look at a measurement process, can provide a mechanism for better understanding of that process (and even research in some instances). Thus, a serendipitous benefit is an improved understanding of the process.

OPERATION

The recommended mode of operation for a quality circle is shown diagrammatically in Figure 20.1. The mode is the same whether the objective is to identify problems to solve or to solve problems already identified. In the case of both, an activity or a process is studied to decide the most beneficial course of action. This may require study and often will require critical consideration to define the problem area more clearly.

Ordinarily, several aspects of the problem will be identified, in which case the options need to be generated. From the thinking of the circle, a number of ideas will surface and idea generation should be encouraged. Even trivial and loosely related ideas should be considered since these, if not directly useful, may stimulate other ideas, thus broadening the basis for selection and action. The recommended operating procedure when generating ideas is via idea writing which minimizes the overdirection of group thinking by dominant individuals.

The written ideas are collected in an idea box, for example (writer's name not included). These are recirculated for reading by the group. After reading by each member, the writing process may be continued for additional cycles to the point of diminishing returns.

The selection process consists of two activities: categorization and prioritization. Categorization, or clustering, consists of grouping by similarities, i.e., going from "local" to "global." This is by way of discussion in which recorded ideas are classified by group consensus (but not rejected). During the discussion, the ideas may be clarified and edited as needed. The process can be carried out by intercomparison, i.e., to distinguish between similarities and differences of ideas, on a one-to-one basis. This process will result in several groups (clusters) each having similar characteristics, the general characteristics of which can be identified. While all clusters will have merit, some will be perceived to be more urgent (or important) than others. Prioritization can then be established by group preferential rating.

The actions that could be taken will fall into three classes:

Iteration – further study may be required to define a problem more clearly for specific action; the problem may need collaborative effort with other circles/groups.

Implementation – the circle may have the authority to act on some of its own recommendations with little or no approval required by others.

Recommendation – most problem solving will fall in this class in that the solution may need approval by management or interaction with others, which may require external approval. In such cases, the circle will decide on the best course of action to follow in presenting its recommendations to management.

REPORTING

Recommendations may go to management by two routes. Written reports provide a record of a circle's outputs and are always needed. In addition, oral presentations may be made. The circle should be well-prepared in such a situation, calling for dry runs. If a dry run does not appear to go well, it may mean that something is lacking in the subject matter. Hence, oral presenta-

Table 20.1. Quality Circle Discussion Topics

Topic
Program/Project Management
Problem identification
Defining project objectives
Experimental design
Defining appropriate models
Factorial design
Project Organization
Quality Assurance Program Development
General aspects
Chain of custody
Record keeping
QA responsibilities
Reporting procedures
Quality Control Techniques
Methods of sample preparation
Contamination control
Analytical factors influencing data quality
Control charts
Quality Assessment Techniques
Quality assessment samples
SRMs
Collaborative tests and test results
Presentation/plotting data
Inspection for quality
Performance audits
Systems audits
Statistical Techniques
Precision/accuracy concepts
Regression techniques
Fitting of equations to data
Statistical tests
Analysis of variance
Statistical reporting of data
General
Nomenclature
Definitions
Critical introspection
Sampling
Maintenance of equipment
Housekeeping practices
Reviews of current books/publications in field
Safety
Basic laboratory precautions
Chemical hazards
Physical hazards
Waste disposal
Storage of chemicals/equipment
Space utilization

tions, whether or not used as a final means of communication, are useful for deciding the merits of any recommendations made by a circle.

AGENDA

Quality circles may investigate problems as a result of their own initiative or as the result of the findings of an audit. A list of suggested topics to broaden understanding of the measurement process is listed in Table 20.1.

Quality circles can provide an excellent mechanism to develop the GLP and GMP documents needed for an appropriate quality assurance program. Many of the items listed in Table 20.1 are excellent subjects for GLPs and GMPs.

Quality circle contributions can range from the identification of subjects, the outlining of the contents, to the writing and/or reviewing of the documents. Documents so developed should have a high degree of credibility and acceptability in a laboratory.

CHAPTER 21

Validation

Validation is the process of determining the suitability of a measurement system for providing useful analytical data [135]. While the term is used most frequently in discussing methodology, it applies to all aspects of the system and especially to samples, their measurement, and the actual data output.

VALIDATION OF SAMPLES

Sample validation has several purposes:

- to accept an individual sample as a member of a
- population under study
- to admit samples to the measurement process
- to minimize later questions on sample authenticity
- to provide an opportunity for resampling when needed

Sample validation should be based on objective criteria to eliminate subjective decisions. Generic criteria for acceptance include:

- positive identification
- conformance with physical/chemical specifications
- a valid chain of custody

Sample rejection can be based on: knowledge that a sampling system was not in control at the time a sample was obtained; erroneous or conflicting data on the identity or character of a sample; questions about a sample that cannot be resolved; or any information that would cast doubt on the status of a sample as a member of the population of interest.

VALIDATION OF METHODOLOGY

Validation of methodology is a value judgment in which the performance parameters of the method are compared with the requirements for the analytical data. Obviously, a method that is valid in one situation could be invalid in another. Accordingly, the establishment of firm requirements for the data is a prerequisite for method selection and validation. When data requirements are ill-considered, analytical measurement can be unnecessarily expensive if the method chosen is more accurate than required, inadequate if the method is less accurate than required, or utterly futile if the accuracy of the method is unknown.

Several approaches may be used to validate methodology. When reference samples are available that are similar in all respects to the test samples, the process is very simple. It consists of analyzing a sufficient number of reference samples and comparing the results to the expected or certified values. Before or during such an exercise, the analyst must demonstrate the attainment of a state of statistical control of the measurement system so that the results can be relied upon as representative of those expected when using the methodology measurement system.

When a suitable reference material is not available, several other approaches are possible. One consists of comparing the results of the candidate method with those of another method known to be applicable and reliable, but not useful in the present situation because of cost, unavailability of personnel or equipment, or other reasons. Even the agreement of results with those obtained using any additional independent method can provide some useful information.

Spiked samples and surrogates may be used as reference samples. This approach is less desirable and less satisfactory because of the difficulty in the reliable preparation of such samples and because artificially added materials such as spikes and surrogates may exhibit matrix effects differing from those of natural samples. Split samples of the actual test samples may be used to evaluate the precision of a method or procedure, but they provide no information about the presence or magnitude of any measurement bias.

Another approach is to infer the appropriateness of methodology from measurements on analogous reference materials. The critical professional judgment of the analyst is necessary to decide the validity of the inference.

In all cases, sufficient tests must be made to evaluate the methodology for the variety of matrices and ranges of composition expected during the measurement process. Ordinarily, the latter should include three levels of concentration—namely, the extremes and the midrange of the compositions expected. Statistical considerations suggest that at least six degrees of freedom (ordinarily seven measurements) should be involved at each decision point.

In regulatory programs, agencies setting legal limits have the responsibility for specifying valid methodology for enforcement; otherwise, nothing but chaos would result from conflicting claims for the methodology used by the regulator and the regulatee. Not only should the methodology be appropriate, but it should be suitable for use under field or other conditions by laboratories believed to be typical users. Accordingly, the validation is best accomplished by collaborative testing using typical laboratories.

An individual analyst or laboratory providing services to others has the responsibility to use methods of demonstrated validity. This includes a demonstration by the individual practitioner of the ability to achieve performance within requisite limits.

VALIDATION OF DATA

Data validation is the process by which data are filtered and accepted or rejected based on a set of criteria. It is the final step that should be taken before release of data. The first requirement is the documented validity of the sample(s) and the measurement process used. Then, statistically supported limits of uncertainty should be estimated. Beyond this, checks should be made to eliminate blunders to the extent possible. These include:

- checks for proper identification
- checks for transmittal errors
- checks for internal consistency
- checks for temporal and spatial consistency

Techniques to do the above include intercomparisons with similar sample data, checks for reasonableness of values with respect to a priori and/or a posteriori limits, data plots, regression analysis, and tests for outliers. The checks may range from spot checks of randomly selected data to a total data analysis.

Data validation can be facilitated as the analyst is fully informed on the nature of the problem, the end use of the data, and even the expected results. Some people frown on the latter, fearing that such knowledge will unduly influence the results on the present samples. The more knowledge a competent and ethical analyst has, the better job he or she can do.

CHAPTER 22

Reporting Analytical Data

Reports of analysis are the written records of analytical services provided to others. Often, they constitute the only record of extensive effort and have more than transitory value. In such cases, it is obvious that the contents of reports should include all information necessary for understanding what was done and for interpreting the experimental results. In other cases, a brief document is all that is necessary, although full experimental details should be a matter of record elsewhere. The analyst writing a report must decide how much information to convey. It should be remembered that readers of reports frequently are less informed than the writer, and results could be misunderstood and misinterpreted when critical information is omitted.

The format to be used and procedures for review and release of data should be defined in each organization and written as a GLP. The following is given to provide guidance in the development of a data reporting policy and a GLP to define the procedures to be followed.

MINIMUM REQUIREMENTS

Data must be technically sound and defensible before it can be reported. The first requirement for reporting is that all pertinent documentation is available and referenced so that it can be found at any time that it might be needed. The checklist in Table 22.1 identifies the most important documentation.

Most of the items in this checklist are obvious, but the information should be a matter of record. The information base should be adequate in kind and detail to enable a third-party expert to reconstruct what was done. Whenever there are significant gaps in the information base, the hazard of reporting data under such circumstances needs to be considered before it is released. Incidently, the ease with which such information can be retrieved is a measure of the effectiveness of a quality assurance system and is a major consideration in some systems audits.

Table 22.1. Documentation Check List

Sample Documentation	
Sampling Plan	[ ]
Sampling Methodology	[ ]
Sample Identification	[ ]
Location, time, etc.	
Chain of Custody	[ ]
Measurement Documentation	
Methodology	[ ]
Calibration	[ ]
Apparatus	[ ]
Operator	[ ]
Quality Assurance	
Controls	[ ]
Control Charts	[ ]
Demonstration of Statistical Control	[ ]
Data Documentation	
Location	[ ]
Data Reduction Procedure	[ ]

DATA LIMITATIONS

Provided that all essential documentation is verified, the next consideration is the limitations to be assigned to the data. The first of these is the accuracy or at least the precision of the data. Limits of uncertainty (error bars) need to be assigned to every piece of data. These are quantitative estimates of the confidence believed by the producer of the data; they indicate the limitations on its use and should discourage its overinterpretation. No data should be released until statistically supported limits can be and are reported for it.

There should be virtually no uncertainty about the identity of what parameter was measured. This is often based on knowledge of the measurement process and the nature of the samples measured. In some cases, and especially in the initial stages of a measurement program, this may need to be confirmed experimentally. Measurement by several independent techniques and variation of experimental conditions are approaches commonly used for this purpose.

The question of recovery is often important. This refers to whether or not the methodology measures all of the analyte that is contained in the sample. This is best evaluated by the measurement of reference materials or other samples of known composition. In their absence, spikes or surrogates may be added to the sample matrix. The recovery is often stated as the percentage measured with respect to what was added. Complete recovery (100%) is of course the ultimate goal. At the minimum, recoveries should be constant (only varying within acceptable limits), and should not differ significantly from an acceptable value. This means that control charts or some other means should be used for verification. Significantly low recoveries should be pointed out, and any corrections made for recovery should be stated explicitly.

The limit of detection (LOD) is of special interest when analytes are sought but not found. Because the LOD is a statistical parameter, it must be based on sound statistical data. Since there are a number of ways in which the LOD can be calculated, the one used must be identified.

NUMERICAL VALUES

The uncertainty of a measured value can be defined by a statistical confidence interval (see Chapter 4) and an estimate of the bounds for systematic error. These should be stated separately. The level of confidence chosen should be consistent in a report, and considerable advantage is gained by adopting a uniform policy for a laboratory. The level chosen must be stated whenever plus-or-minus values are reported. A 95% level is reasonably conservative and used by many investigators.

In some cases, a statistical tolerance interval may be reported (see Chapter 4), representing the range within which a specified percentage of the individual values for a population (measurements or samples) is expected to lie with a stated level of confidence. A uniform convention in reporting is again advised, and the 95%/95% (95% confidence that 95% of the population is included in the interval) is recommended. Again the description of the interval must be stated explicitly.

In both of the above cases, there must be no confusion as to the population (measurements or samples) to which the values apply.

LIMITS OF UNCERTAINTY FOR DATA

When the measurement process is demonstrated to be in a state of statistical control, the process standard deviation may be used to evaluate the confidence interval for the mean of n measurements. The use of control charts is the best way to demonstrate that a process is in a state of statistical control at the time the measurements are made, and this can minimize the amount of work needed to establish confidence limits for the data.

In the absence of a control chart, a sufficient number of replicate measurements must be made to demonstrate statistical control and to estimate the standard deviation of measurement with a reasonable degree of confidence. The minimum number of replicate measurements required is somewhat arbitrary and depends on the risk concerned with exceeding the limits of confidence that are stated. Metrologists often recommended 7 to 30 determinations as a reasonable number of replicates. The uncertainty of the standard deviation increases rapidly below 7, and little is to be gained by increasing the number above 30.

The question of possible bias must always concern the analyst. It is not considered good practice to correct for biases without understanding their origin; hence, every reasonable effort must be made to identify and quantify potential contributors. Biases such as analytical blanks should be measured as accurately as necessary and possible, and the results are corrected directly for them. This is proper because analytical chemists believe that blanks are additive errors. Biases such as those found when measuring SRMs should be investigated to identify their source so that they can be eliminated, minimized, or corrected for in a proper manner. As already pointed out, the method of correction will be dictated by the nature of the bias. The most reliable approach is to remove the bias rather than to correct for it.

The treatment of bias related to the question of recovery often troubles the trace analyst. Usually, corrections are not made, but recoveries are reported as one of the qualifications for the data. The recovery, no matter what its value, should be shown to be in a state of statistical control, and this should be a requirement for reporting data of this kind. When a recovery determination is made and the value obtained is variable or does not fall within control limits, corrective action is indicated, and control should be reestablished before data may be reported.

One can make an analysis of bias as one would make an analysis of variance. First, all sources of bias are identified as possible—that is to say, a bias budget is developed. Although corrections may have been made for them, the uncertainty of such corrections should be estimated. These are then summed, algebraically, to obtain the total bounds for bias based on professional judgment. The basis for such judgment should be a matter of record.

There are several schools of thought on how to combine statistical confidence limits and the bounds for bias. The most conservative approach is to add the two estimates algebraically. This approach is based on the concept that biases are not necessarily randomly distributed and could all add in the same direction. This approach is used when certifying SRMs at NBS and is the one recommended. On the other hand, some experimenters add both kinds of error in quadrature, which may be called the most optimistic approach. Because there are several ways to report data, the approach used should be clearly described.

SIGNIFICANT FIGURES

Numerical data are often obtained (or at least calculations can be made) with more digits than are justified by its accuracy or precision. So that it is not misleading, such data, when reported, should be rounded to the number of figures consistent with the confidence that can be placed in it. Accordingly, metrologists have adopted the terminology of significant figures in describing the resulting data. The number of significant figures is said to be the number of digits remaining when the data is so rounded. The last digit, or at most the last two digits, are expected to be the only ones that would be subject to change on further experimentation, for example. Thus, for a measured value of 20.5, only the 5, and at most the 0.5, would be expected to be subject to change. Such data would be described as having three significant figures.

In counting significant figures, any zeros used to locate a decimal point are not considered as significant. Thus 0.0025 contains only two significant figures. Any zeros to the right of the digits are considered as significant; therefore, only those that have significance should be retained. Thus, 2500 and 2501 each have four significant figures. Zeros should not be added to the right of significant digits to define the magnitude of a value unless they are significant, since this would confuse the significance of the value. For example, it is not good practice to report a value as 2500 ng, but rather 2.5 μg, if the data is reliable to two significant figures. The use of exponential notation, e.g., 3.5×10^3, is an acceptable way to express both the number of significant figures and the magnitude of a result. The concept of significant figures is especially important in mathematical calculations. In multiplication and division, the operator with the least number of significant figures determines the numbers to be reported in the result. For example, the product $1256 \times 12.2 = 15323.2$ is reported as 1.53×10^4. In addition and subtraction, the least number of figures to either the right or the left of the decimal point determines the number of significant figures to be reported. Thus the sum of $120.05 + 10.1 + 56.323 = 186.473$ is reported as 186.5, because 10.1 defines the reporting level. In complex calculations involving multiplications and additions, for example, the operation is done serially, and the final result is rounded according to the least number of significant figures involved. Thus, $(1256 \times 12.2) + 125 = 1.53 \times 10^4 + 125 = 1.54 \times 10^4$.

The following rules should be used in rounding data, consistent with its significance:

1. When the digit immediately after the one to be retained is less than five, the retained figure is kept unchanged. For example: 2.541 becomes 2.5 to two significant figures.
2. When the digit immediately after the one to be retained is greater than five, the retained figure is increased by one. For example: 2.453 becomes 2.5 to two significant figures.
3. When the digit immediately after the one to be retained is exactly five and the retained digit is even, it is left unchanged and conversely. For example: 3.450 becomes 3.4, but 3.550 becomes 3.6 to two significant figures.
4. When two or more figures are to the right of the last figure to be retained, they are considered as a group in rounding decisions. Thus in 2.4(501), the group (501) is considered to be >5 while for 2.5(499), (499) is considered to be <5.

GUIDELINES FOR REPORTING RESULTS OF MEASUREMENTS

The number of significant figures to be used in reporting results is often a question. This will depend on the number of figures in the original data and the confidence limits to support the results. Analysts sometimes feel that observed data have more digits than are meaningful and are tempted to round them to what is felt to be significant. This should be resisted, and rounding should be deferred as the last operation. The following guidelines are recommended when deciding what is significant.

The number of figures to retain in experimental raw data and even in preliminary calculations is unimportant, provided a certain minimum is exceeded. At least the last figure should vary between successive trials, and variability of at least the last two figures is preferred. If this is not the case, the data are probably truncated by the operation (e.g., low attenuation), rounded off by the observer, or imprecisely read. Training of observers can often improve the precision of reading. Observers can have preconceived ideas of the attainable precision (or that required for some application) and will round off readings with this in mind. Thus, they may be actually throwing away data.

The average of several values should be calculated with at least one more figure than that of the data. This will then be rounded for reporting, consistent with the confidence limits estimated.

The standard deviation (necessary for computing confidence intervals) should be computed to three significant figures and rounded to two when reported as data. The confidence interval should be calculated, then rounded to two significant figures (use more than this number in the calculations if available), and the result reported should be consistent with this. While any confidence level may be used, the 95% level is commonly used. However, the level used for the calculation must be reported. As an example, the following data were observed:

$$15.2,\ 14.7,\ 15.1,\ 15.0,\ 15.3,\ 15.2,\ 14.9$$

$$\bar{X} = 15.057 \qquad n = 7$$

$$s = 0.207 \qquad df = 6$$

Confidence interval calculation:

$$\frac{t\,s}{\sqrt{n}} = \frac{2.517 \times 0.207}{\sqrt{7}} = 0.1969$$

The result reported is $\bar{X} = 15.06 \pm 0.20$ where the uncertainty represents the 95% confidence interval for the mean of seven measurements.

GUIDELINES FOR COMBINING DATA SETS

There are various occasions when it is desired to combine sets of data to obtain a broad-based estimate of a measured property. The rationale for such a combination is outlined in Figure 22.1. Only compatible data should be combined. Any data that are not supported by confidence limits are rejected as "quality unknown." The remaining data are screened for the significance of apparent differences. For large numbers of data, this may be done by looking for outliers. Whenever such are found, an assignable cause should be sought, and effort should be made to salvage rejected data. If this is not possible and there is no reason for selection, it may be necessary to reject all data or to report the several results independently.

The means of sets that can be combined are calculated on the basis of one of the following assumptions:

- All values are based on the same number of measurements, all of equal precision.
- Values are based on different numbers of measurements, but there is no reason to believe that the precisions differ.
- Values are based on different numbers of measurements and/or with different precisions.

The statistical procedures applicable to each of these situations are described in Chapter 4.

ARBITRARY REPORTING PROCEDURES

Some laboratories follow arbitrary rules when reporting data, based on preconceived ideas of its quality and/or on the needs of the user. These may or may not be statistically supported. Data formats for computer storage may influence how data is reported. Whenever data are truncated, information which could have value may be lost. For example, small differences may be masked. Accordingly, arbitrary reporting practices should be carefully considered. As a minimum, statistical computations should precede any rounding that is done on the data reported.

REPORTING LOW-LEVEL DATA

Several approaches have been recommended for reporting low-level data. The American Chemical Society, Committee on Environmental Improvement, recognizing the problems of misuse of data having relatively high uncertainty,

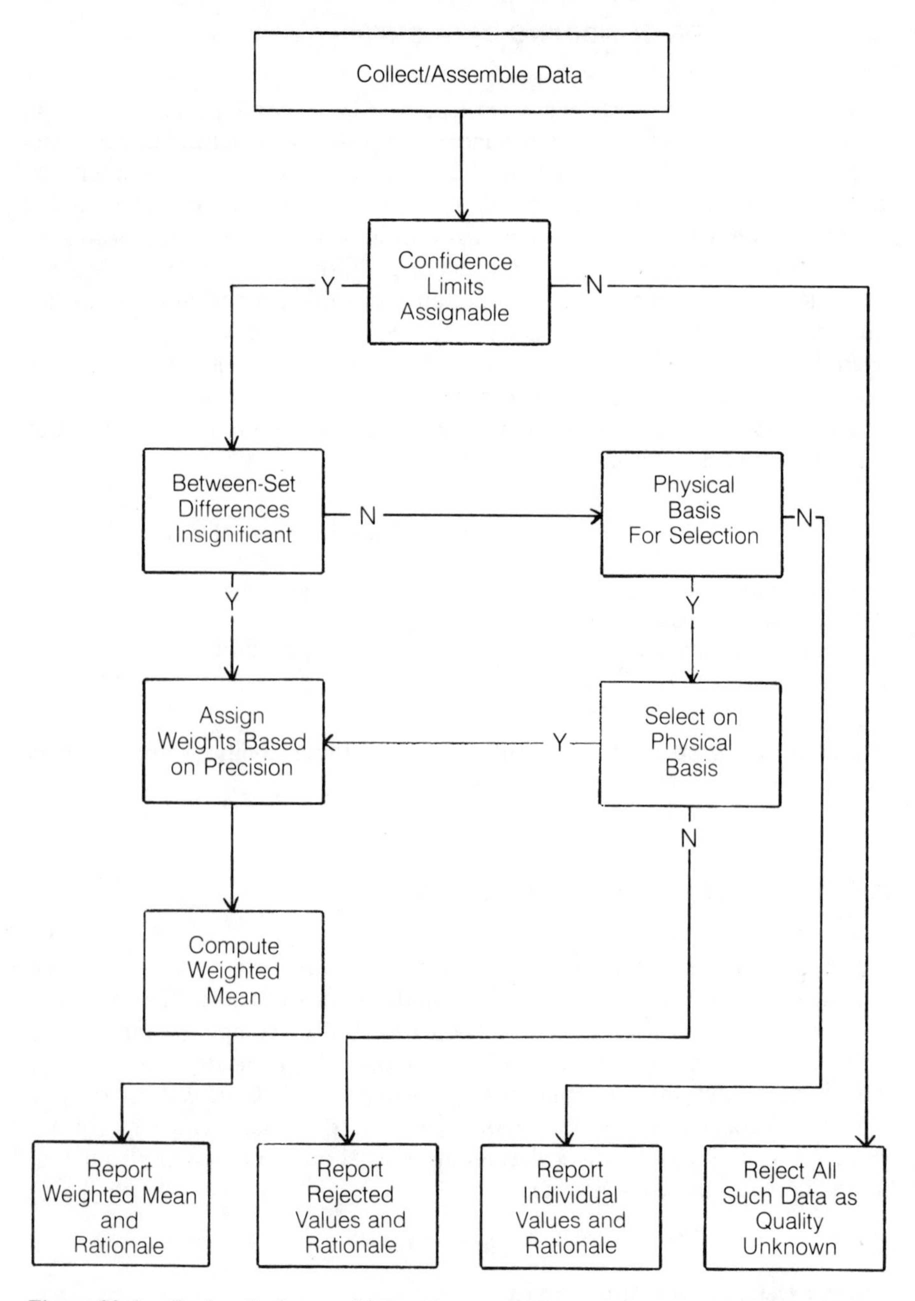

Figure 22.1. Rationale for combining data sets.

recommends the following based on a knowledge of this, the estimated standard deviation of measurement near the LOD:

Guidelines for Reporting Data [2,3]

Analyte concentration in units of s	Report
$< 3s_0$	not detected
$3s_0$	limit of detection
$3s_0$ to $10s_0$	found but not quantitated
greater than $10s_0$	numical value reported with associated uncertainty

ASTM Standard D-4210

The ASTM Standard D-4210 [12] recommends reporting numerical values whenever found, together with their associated uncertainties. It recommends reporting negative values when found (such as would be obtained when subtracting background from signal or blank from sample).

The ACS guidelines are most appropriate when interpreting individual values, while the ASTM approach is necessary when a number of data values are to be combined.

REPORT FORMAT

When routine data are reported, a simple table may be all that is necessary. Together with or supplemental to this should be an explanation of the significance of the values reported. In some cases, a coding system is used to indicate limits of uncertainty. For more extensive data, and especially for data that is expected to be used for nonroutine purposes, a more or less detailed report may be merited. The following format may be used for guidance in preparation of a comprehensive report.

Content of Report of Analysis

The following outline has been prepared to provide guidance in drafting a report of analysis that will be informative to the client and provide historical documentation as it might be needed for future reference. Another objective is to facilitate uniformity in reporting, thus enhancing communication. Not all items will be pertinent in every case, but all should be considered each time a report is written. Items should be considered from the viewpoint of a reader with general but not specific knowledge in the field of analysis. Above all, the report should be informative to the client and should describe the data ,including their strengths and weaknesses.

Title. Brief but descriptive title to identify the work done and facilitate information retrieval.

Client. Full identification of the sponsor for whom the work was done. This may include: name, organization, address, purchase order, work order, cost center, etc., as appropriate.

Report number. Number used to identify the report. This number should appear on each page, including figures and tables. A suggested form is: Calendar year/Sequence No. The sequence number is assigned by the laboratory management and should be entered in a logbook with the report title.

Date. Date report is completed.

Objective. Brief statement of the work to be done. This can be taken from the client's request or from other sources. The reason why the work was done, any assumptions on which it is based, and the way the data are expected to be used may be included.

Sample identification. A clear but definite description of the material analyzed including any known limitations that would impact on the data. Examples include: serial number, analyst's description of samples, photographs with sampling area clearly indicated, and physical description of sample (e.g., single phase, heterogeneous powder, etc.).

Sampling details. How the actual samples analyzed were obtained, including:

- sampling procedure
- use of an entire sample or subsample
- subsampling procedure
- homogenization procedure
- sample chemical/physical treatment
- type of sample
- random sample by client
- random sample based on statistical protocol
- selected sample (basis)
- chain of custody (as appropriate)
- disposition of sample after analysis (as appropriate)

Summary of or reference to analytical method used. Brief but informative account of how measurements were made. This can be a reference to a published method or to one in use within the organization. When a reference is given, be sure that it is specific, reliable, and provides the information intended. Any modification of a referenced method should be stated explicitly.

Calibration, standardization, and controls. General statement of basis for calibration and standardization. (Details should be with original data and need not be included here if documented elsewhere.) Statement of controls used such as measurement of an SRM and inferences of statistical control derived from such measurements. Basis for assignment of uncertainty (precision, bias) should be stated, if not from data set.

Results and limitations. Clear summary of data with explicit statement of uncertainty. Uncertainty relates to:

- measurement process
- sample homogeneity
- between sample variance
- relation of data to population from which sample was taken

Uncertainty also relates to any other limitations on the data that would impact on its interpretation or use. An example is its restriction exclusively to the sample(s) measured when this is the case.

Discussion. Relation of results obtained to the objective of the analysis. This can include analyst's interpretation of the results. Recommendations for additional work needed to serve the objective or to clarify questions raised during the measurements may be included. Any other information, not included elsewhere, may be presented as pertinent to the analysis or matters related to it.

References. References to notebooks, charts, data printout, etc., where raw data and full experimental details are recorded or documented. (These may be on blind copy not given to client.) Cross references to other reports may be given as pertinent. These may include previous work (in a series) or work done by others. If the present work depends on other work, this must be stated.

Attestation. Signatures and titles of all who have responsibility for the content of the report and/or authority for its release should be included. This will follow organization policy. In some cases, release is contingent on approval at higher management levels.

Distribution list. The full distribution list for the report should be listed for reference as necessary.

Miscellaneous. Pages should read "page x of y" so that clients are certain they have the full report.

CHAPTER 23

An Improved Approach to Performance Testing

Performance testing (PT) is a special kind of collaborative test or round robin [142] conducted to evaluate the performance of a laboratory for reasons such as the following:

- to test the ability of a laboratory to produce data of a specified quality as might be required by a regulatory process to qualify laboratories as peers in a monitoring program when data must be compatible
- to meet requirements for certification
- to meet qualifications for contractual purposes
- as a means for a group of laboratories within a peer group to demonstrate peer performance
- as a means for a laboratory voluntarily to assess its own performance

Although widely used, performance testing as ordinarily conducted is often a poor means to serve the purposes given above. Usually, a limited number of measurements are made on a small number of samples by a few participants. Accordingly, the statistical power is low so that only gross problems can be identified that should be recognizable even without statistical analysis. At best, only performance of the moment is evaluated with no guarantee that it is a representative sample of ongoing performance, thus raising questions as to what is really tested.

If a laboratory were required to establish statistical control and to evaluate its own precision prior to participating in a PT, a few PT measurements could be very instructive in performance evaluation. Even the proficiency of a lone participant can be evaluated. The PT then becomes a way to confirm a laboratory's self-evaluated precision and to identify measurement bias. Accordingly, the PT procedure is simplified, and the results can be interpreted much better than in the case of a conventional PT.

PREMISES

Ordinarily, a laboratory participates in a PT because it already does or desires to offer services that need to be evaluated. As a provider of such (even for internal consumption), it has the obligation to have acquired at least a minimum level of expertise with the appropriate method of test and to have evaluated its own precision. After all, this is a basic requirement—attainment of statistical control—before there can be any confidence that anything at all has been measured.

The purpose of a PT is to evaluate expertise, not to gain expertise. Accordingly, participating laboratories should be required to provide documented evidence of statistical control (preferably by means of a control chart) and of the precision they have attained.

As a further requirement, each participating laboratory should have a documented procedure (SOP) for the measurements to be made. In some cases, this SOP may be reviewed and approved by the organizer of the PT for its apparent adequacy or equivalency to a specified method. When submitting their results, laboratories should certify that they have followed the SOP precisely or describe any deviations from the procedure. A copy of the SOP should be made available to whomever critiques the test results.

Furthermore, it is recommended that each participating laboratory should have an operational quality assurance program, and the PT should be conducted under a protocol (PSP) that defines the good laboratory practices (GLPs) and the good measurement practices (GMPs) that are pertinent to it. In monitoring programs, the PSP should be the subject of critical review prior to participation in the PT.

Before the PT is conducted, there should be some expectation of what would be classed as good performance. Precisionwise, this should not be a problem since this can be obtained from the experience of a reference laboratory, from the collaborative test data (or validation studies) when a standard or reference method is used, from the pooled standard deviations submitted by the participants, or from requirements for usable data.

Only peer laboratories (on the basis of their documented precision) should be permitted to participate. That is to say, if a laboratory has not achieved peer status (based on its own self-evaluation), little will be learned from its participation in a PT, but rather it should devote its efforts to such attainment. Otherwise, the occurrence of outliers is expected. There should be no outliers in a peer group (as defined above).

ADVANCE ESTIMATION OF PRECISION

Advance estimates of precision should be based on no less than 7 and preferably at least 15 degrees of freedom. While a laboratory should know both its short-term (s_w) and long-term (s_b) standard deviations, the latter is the one of concern in performance testing. It is recommended that precision estimates be based on duplicate measurements made on a suitable number of typical samples with no two sets of duplicate measurements made on the same day. Such a measurement program will permit evaluation of both short-term and long-term precision. (See section at the end of this chapter for the statistical treatment of such data.) The latter is the parameter most useful for computing confidence limits for reported data and often exceeds the former by a factor of 1.5 to 3, even in a well-operating laboratory.

PERFORMANCE TESTING PLAN

The following outlines the recommended basic steps of a PT exercise:

1. Each participant must meet peer qualifications as already discussed.
2. A decision is made in advance on what evidence of statistical control will be required. If a control chart is a requirement, control samples identical to those of the chart should be required to be analyzed during the course of the PT.
3. A set of essentially homogeneous samples is sent to each participant, preferably at three levels spanning the range of interest (and preferably in duplicate).
4. Each participant makes duplicate measurements on each sample and on control samples, as appropriate.
5. Results of each laboratory are statistically analyzed (hypothesis testing) for evaluation of individual bias. In this analysis, the laboratory's own precision, the pooled standard deviation of the participants, and the pooled standard deviation of the PT are used to make separate estimates of bias.
6. In the case of multiple laboratories, the results are screened for outliers, after which a between-laboratory standard deviation is calculated. If this is substantially different from the pooled long-term standard deviation of the participants, the PT results should be examined critically for sources of bias and particularly for calibration problems.

DIAGNOSIS

The critical parameters of each test method should be known and tolerances evaluated for them to the extent possible before a PT is undertaken. Ruggedness testing is recommended for this purpose[155]. It is difficult if not impossible for a PT to investigate this. However, it is possible to include a "normalization" sample to give a common reference basis for the test results, especially when trying to minimize biases due to calibration, for example. When normalization significantly improves between-laboratory agreement, calibration problems should be looked for and corrected. The so-called between-laboratory standard deviation is often misleading since it is influenced by interlaboratory biases that are not randomly distributed and may even be artifacts of a particular PT. Moreover, the acceptance of such lends dignity to the culprit that the PT aims to detect and abolish—namely interlaboratory bias.

In any consideration of bias, one needs to distinguish between method-inherent and application-related sources. The former should be consistent for every participant while the latter will vary with participants and even from time to time. Method-inherent bias requires research for its correction. Elimination of application-related bias requires improved quality assurance practices.

The analysis for outliers is especially important. While they should be eliminated for the purpose of statistical analysis, they should not be forgotten, but rather "assignable causes" should be sought for anomalous performance. After all, each laboratory was originally qualified as a peer; hence, "assignable causes" are of interest in identifying measurement problems and improving the future performance of all laboratories. If the reason for rejection cannot be found, it must be blamed on a "blunder" or on a defect of the methodology. Clearly, research should be undertaken if the latter is the case.

FEEDBACK

Evaluation of results is an essential part of performance testing. Each participant should be informed of its performance in relation to an acceptable standard. If the PT is directed toward the evaluation of a group of laboratories, a summary of test results should be circulated to each participant, as a minimum. Coding of results to ensure anonymity is recommended.

Individual laboratories should be requested to analyze their own results, seeking the assistance of the organizers as necessary, and to take corrective actions to their measurement systems. By comparison of its test results with its control charts or other evidence of internal performance, each laboratory should be able to identify the nature of any defective performance, and the remedial actions required usually will be clear. Group meetings in which participants discuss their results is an excellent way to share experiences to correct measurement problems.

It is illogical to expect that problems will disappear without corrective actions. Accordingly, each participant should consider its performance critically, take appropriate corrective actions, and document what was done and the reasons for doing so. If no actions are deemed necessary, even this is an action that should be justified. In continuing PT exercises, the preparation of a self-evaluation report and the implementation of corrective measures should be a prerequisite for participation in succeeding exercises.

Test results must not be compromised by uncertainty in the test samples. High-quality test materials are an essential requirement for any PT (see Chapter 16). The development and prior evaluation of the test materials can be a critical aspect of a PT program. Moreover, the test must be designed to forestall any question about adequate samples or to identify sample uncertainties if they exist (see Chapter 16).

CONSENSUS VALUES

In the absence of the "true value" for a test material, a consensus of results is sometimes used to evaluate laboratory bias. Likewise, such results may be used to assign a value to a reference material. Accordingly, a few general comments on the merits of such an approach are in order.

In the absence of bias, the differences of results reported by collaborators are caused by imprecision of measurement. The consensus value, when the results of each collaborator are properly weighted (see Chapter 4), should be a better estimate of the value of the property of concern than any single measurement.

When bias is or may be present, the results of the collaborators may be influenced by it as well as by imprecision. In such a case, the consensus results have little significance unless biases of the laboratories are randomly distributed. Only in the case of a large number of collaborators can such an assumption be justified. However, such consensus values can always be questioned, and the experimental proof of random distribution may be difficult to achieve. It is not uncommon that an apparent outlier is closer to the "true value" than the consensus value due to biases of the majority.

The consensus values resulting from a small number of contributors can have significant uncertainty, even in the absence of bias. When bias can be present, such a value may have limited usefulness.

STATISTICAL TREATMENT OF DATA

For each measurement level, list data as follows:

Lab No.	X_A	X_B	$R = \lvert X_A - X_B \rvert$	$\bar{X}$
1				
2				
.				
.				
k				
			$\bar{R} =$	$\bar{\bar{X}} =$

1. Calculate $\bar{R} = \dfrac{R_1 + R_2 + \ldots + R_k}{k}$

2. Calculate $s_w = \bar{R}d^*_2$

   The value d^*_2 will depend on k (also on number of replicates). See Table C.1.

3. Calculate $s_x = \sqrt{(\frac{\Sigma(\bar{X}_i - \bar{\bar{X}})_2}{k - 1})}$

4. Caculate $s_b = \sqrt{(s^2_x - s^2_b/2)}$

5. Calculate s_b (pooled) from the values for each laboratory, as described on page 213.

Notes

1. Examination of R values.

   R is a measure of the standard deviation. For a system in control, any value for R should not differ from $\bar{R}$ by more than a factor of 3.3 (at 99.7% confidence level). If a larger value of R is obtained by any laboratory, note whether X_A or X_B or both differ appreciably from those reported by others. If one differs, look for a problem with that value or with the sample. If both differ, look for a measurement problem.
2. s_b should compare favorably with s_b (pooled) computed from the values reported by the individual laboratories. If not, look for outliers in the x values due to biased measurements.
3. Further information on statistical evaluation of collaborative test data is available elsewhere [18,159].

CHAPTER 24

Correction of Errors and Improving Accuracy

Analytical errors and undesirable performance are discovered as the result of a laboratory's own quality assessment efforts, from its participation in collaborative test programs, or from external performance audits. The following discussion deals with the correction of problems and improving the specific performance of a laboratory.

IDENTIFICATION OF PROBLEMS

The first objective should be to positively identify the kind of problem that is present, whether due to imprecision or bias. In the event of both possibilities, problems of precision must be addressed first. It is virtually impossible to identify a bias equal or less in magnitude than that of the imprecision. The next factor that should be decided is whether the problem is persistent or transitory. Until these facts are established, it will be difficult to identify the corresponding causes.

There are two kinds of causes that affect a measurement process. Chance causes, sometimes called common causes, occur randomly and affect precision. Ordinarily these are reduced to acceptable levels, and the system is stabilized by using suitable quality control measures to maintain this level of output. However, one should remember that improvement of a measurement process results only from the elimination or reduction of chance causes and their corresponding effects.

Assignable causes, sometimes called special causes, can produce bias or unacceptable imprecision. They may represent malfunction of the quality control procedures, changes in some part of the system, or defects in the system not previously recognized. In fact, as a measurement system is "debugged" and fine-tuned, small measurement problems swamped by more predominant ones emerge and literally beg for their solution as the latter are corrected. It is for this reason that quality assurance has been called a never-ending improvement of quality and productivity [43].

In a manufacturing process, corrective actions often consist of identifying defects, determining their frequency, identifying their source, and making

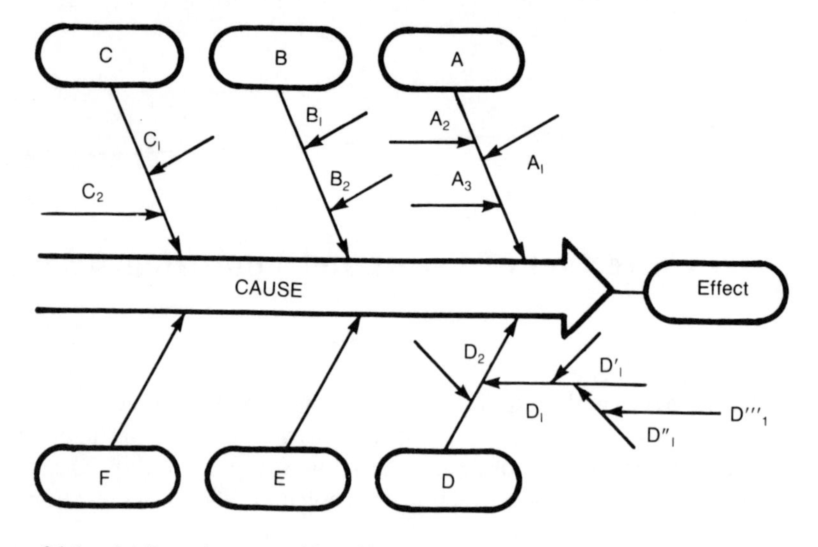

Figure 24.1. Ishikawa's cause-effect diagram

corrective actions. Records are kept of what happens, and the above process is continued until the defects are minimized or eliminated. Pareto analysis based on the frequency of occurrence of specific defects is commonly applied in industrial production situations when deciding the most effective procedure to follow to eliminate them. This approach is possible in measurement situations but is most effective when large numbers of items are involved so that the frequency determinations are reasonably quantitative.

The identification of measurement problems is facilitated by using the approach described by Ishikawa [68] based on the use of a cause-effect diagram (sometimes called a fish bone diagram). Such a diagram is shown schematically in Figure 24.1. The idea behind its use is that every effect has a cause or causes that could even be multifaceted. Thus, several variables may be involved, each influenced by subeffects that may be further subdivided. The variables are indicated by the ovals A, B, C, etc. The lines shown for the subcauses are not vectors but are drawn merely for convenience to illustrate the various interactions. If only a few possible causes are identified, either the problem under consideration is a very simple one, or one may not be looking at it very intensely. On the other hand, if too many possible causes are unearthed, it is a very complex problem indeed, or some of them may be subcauses of more major items. The development of cause-effect diagrams for specific measurement situations may be a useful exercise for quality circles. This activity can be a useful teaching tool as well.

The following is another approach for investigating measurement problems. Based on knowledge of the measurement system, a checklist can be developed

Table 24.1. Deficiency Correction Check List

When Measurement Process is: A. In control, but results questioned or questionable B. Out of control Possible Problems	A	B
Procedural changes	*	*
Sampling	*	
Sample handling	*	
Analytical procedure		
Calculations	*	*
Data	*	*
Reagents		*
Equipment		*
Calibration		*
Maintenance		*
Methodology	*	*
Blunder	*	*

that identifies the step or process that may be faulty. Such a checklist is given in Table 24.1. Detailed studies of a specific process can identify the operations involved and the parameters that must be controlled to prevent bias and/or unacceptable imprecision. Such an analysis could be useful in identifying specific problems and also as the basis for establishing tolerances that must be maintained to control the process.

IMPROVEMENT OF PERFORMANCE

Performance evaluation is often based on the results of a collaborative test (see Chapter 23). In such cases, the organizers of a collaborative test or an individual participant may want to know why the test results did not meet expectations. If the confidence limits for the materials analyzed are known and each participant can document its precision, the statistical significance of any apparent differences can be determined. Often, the differences will be significant and caused by biases of one kind or another for which assignable causes should be sought.

One source of bias could be the samples analyzed and that participants received significantly different samples. To answer such questions, Youden [155] suggested that two samples be sent to each participant. It is very unlikely that both samples would be biased in the same way. Accordingly, if the results of the measurements for one sample are plotted with respect to those obtained for the other, the nature of an analytical discrepancy can be identified.

On the basis of random error alone, the results are expected to be distributed

Table 24.2. Directions for Calculations for a Youden Plot

1. Tabulate reported results on Sample A and Sample B for each laboratory
2. Calculate (A–B) for each laboratory and the average (A–B)
3. Calculate R = [(A–B) – (A–B)] for each laboratory
4. Calculate $\bar{R}$ = average of absolute values of R
5. Calculate x = 0.886 $\bar{R}$
6. Calculate the 95% confidence circle radius, r + 2.448 s

in a circular pattern around the expected result (coincidence of the known values for the samples). The dimensions of the circle will depend on the probability level chosen.

When bias is the predominant source of error, the data will be distributed along the 45 line and enclosed in an ellipse, the major axis of which is related to bias while the minor axis is related to precision of measurement. In practice, one rarely gets a perfectly circular distribution when small sets of data are involved. Mandel and Lashoff have considered this question and provide statistical guidance for determining the significance of any apparently elliptical distribution [89].

Table 24.2 contains directions for making the calculations needed for a Youden plot. Figure 24.2 is a Youden plot of the results of duplicate measurements on the same test sample by a group of laboratories. It is clear from the distribution of the results which do not all fall within the precision circle, that within-laboratory biases are responsible for the incompatibility of the measurements.

Measurement at only one level of analyte may not be sufficiently informative on the nature of analytical error. Biases may be constant, level-related, or a combination as shown in Figure 24.3. It is not possible to determine which line, such as A, B, and C, represents points 1 and 2; hence, one cannot know what type of error, if any, might be involved. A minimum of two points will define a line if it is linear and hence identify the type of bias that might be present in the process. Three points are better, using samples in the low, middle, and high range of measurement.

In any test of a measurement process, there is always the danger that the performance characteristics at the time of test are atypical. Accordingly, performance tests are best carried out periodically in connection with an appropriate control chart that is kept by each participant on a continuing basis. In this way, both the short-term and long-term capability of a measurement process can be evaluated.

The discussion above has been concerned with ways to identify measurement problems and to investigate their sources. But the goal of quality assurance is error prevention rather than error detection. Accordingly, laboratories should keep appropriate records of any problems and the corrective actions that are taken. These records should be reviewed on a regular basis for the effectiveness of the actions and also for any patterns or persistence of specific

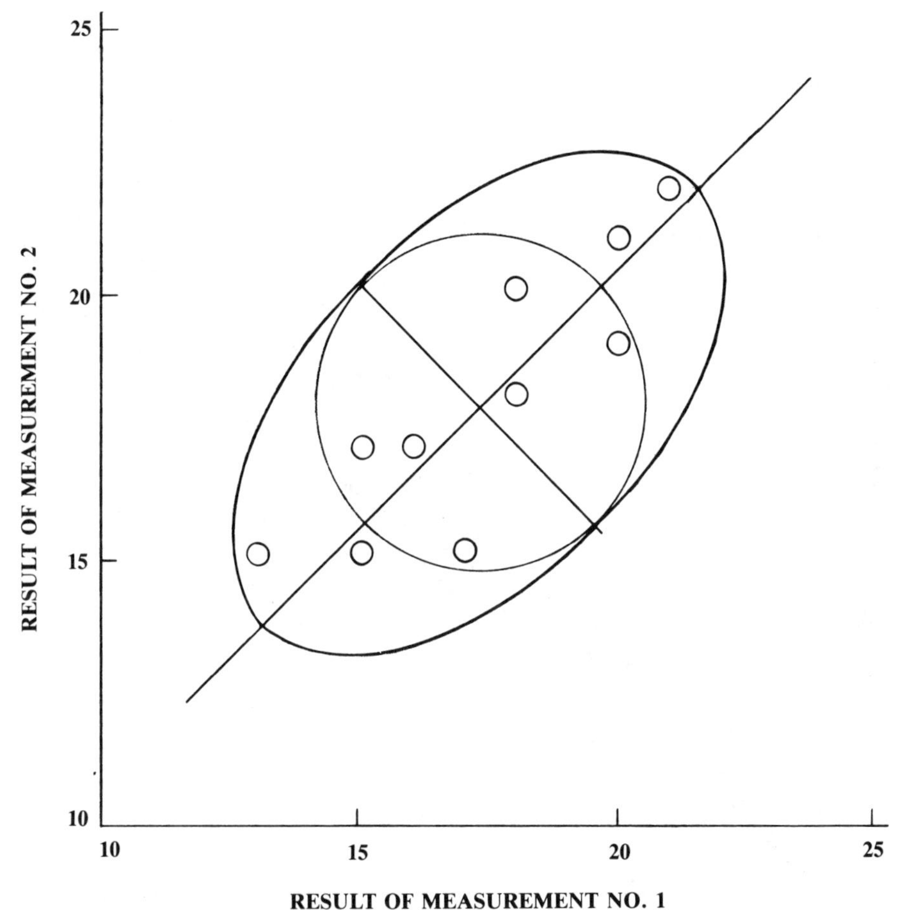

Figure 24.2. Youden plot to identify measurement problems

problems. Pareto analysis referred to earlier may be useful for this purpose. The frequency of occurrence of specific defects or classes of defects is plotted. Obviously, the most benefit would be derived from correcting the problems that occur most frequently. Pareto analysis is most useful when relatively large amounts of information are available and is of little use in the case of rarely occurring problems.

Another way, favored by the author, to analyze measurement problems is by the use of templates as shown in Figures 24.4, 24.5, and 24.6.The idea is that a problem, question, or situation can be modeled by a finite number of major factors which, in turn, may be modeled using subordinate templates. The number 10 is a convenient number to guide thinking in this respect. If you cannot think of ten factors, this may be justified, or you may not be thinking far enough. If you end up with more, you either have a very complex problem, or some factors may be too narrowly defined.

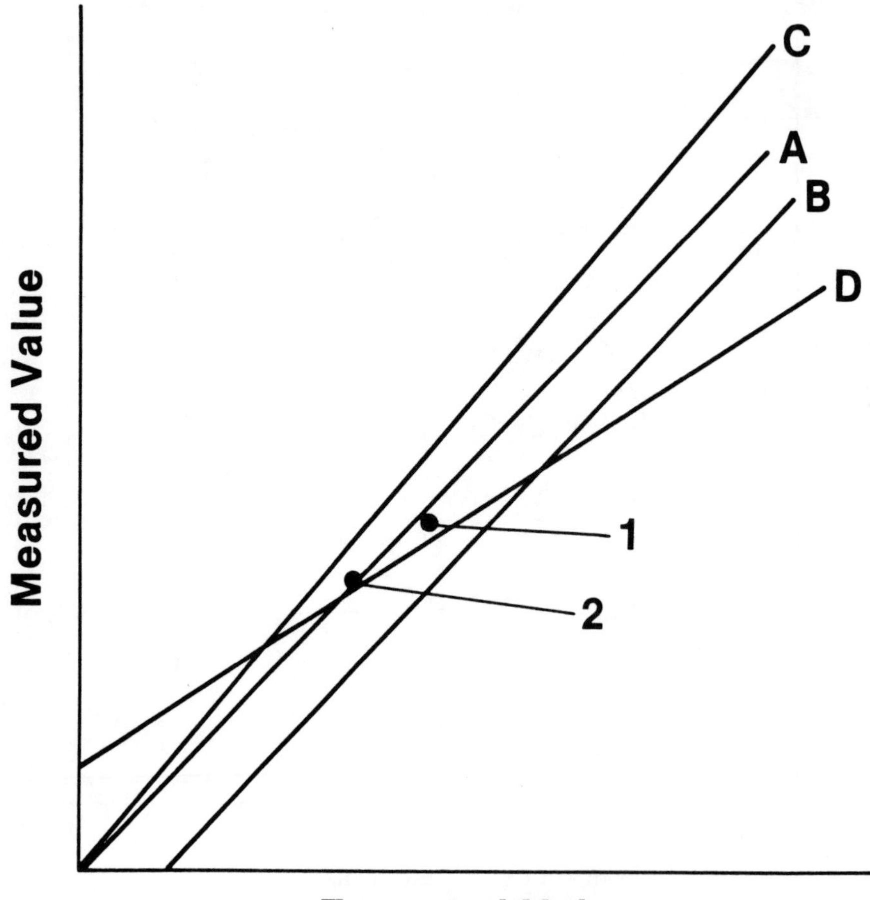

Figure 24.3. Types of analytical bias

As an example, blunders may result from the causes shown in Figure 24.4. You may think of others to fill the blocks. Some of the causes of imprecision and bias are shown in Figure 24.5 and Figure 24.6, respectively. The details related to a specific problem could be inserted in each block. Additional templates with specific causes in the center box (variability of blank, for example) could be used to further define a measurement problem. A blank template that may be duplicated and used in this approach is included in Appendix D.

It cannot be overemphasized that identification of a generic or specific problem is only the beginning step. Any technically significant problem should be investigated for an assignable cause and minimized if not eliminated. The result will be a more useful, reliable, and stable measurement system.

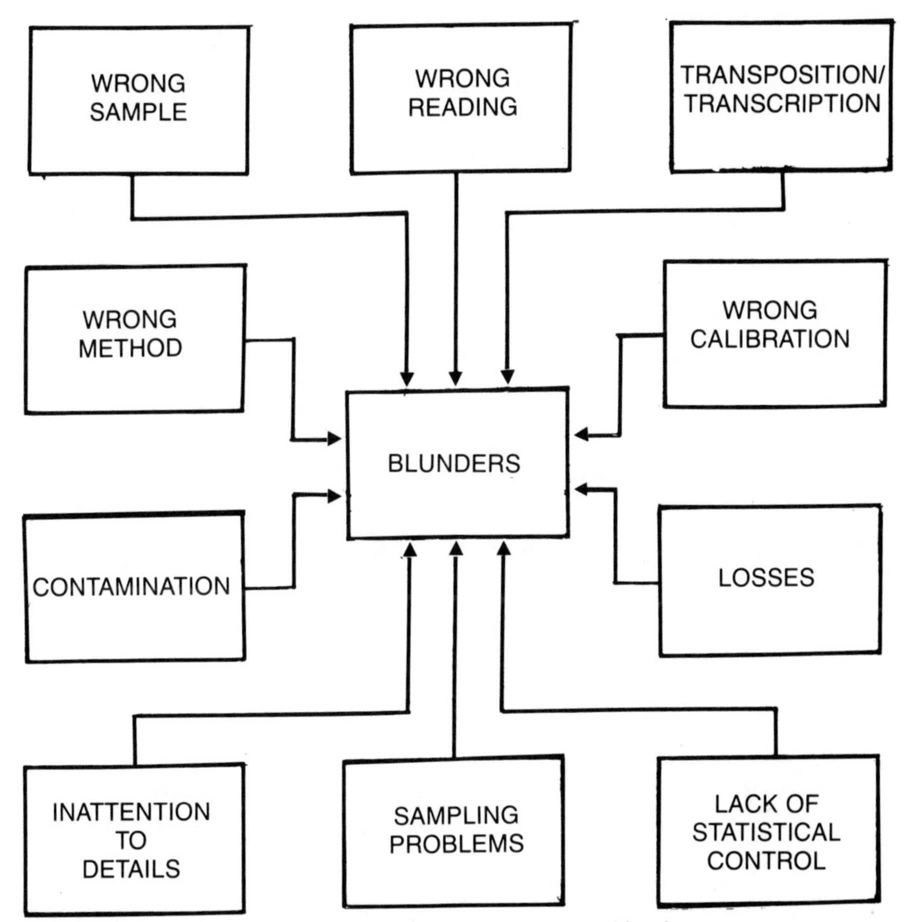

Figure 24.4 Template depicting sources of measurement blunders.

Important as it is, one must remember that the elimination of assignable causes will not improve the stable operational performance characteristics of a measurement system [43]. This can only happen as the result of eliminating or minimizing the effects of common causes. Assignable causes happen only on occasion, while common causes are present all the time until removed. A laboratory should continually examine its measurement processes for improvement, based on a full understanding of all sources of variance and bias.

EDITING DATA

The meticulous analyst can become unduly concerned with the quality of data to the point that every value is critically screened for its credibility. This

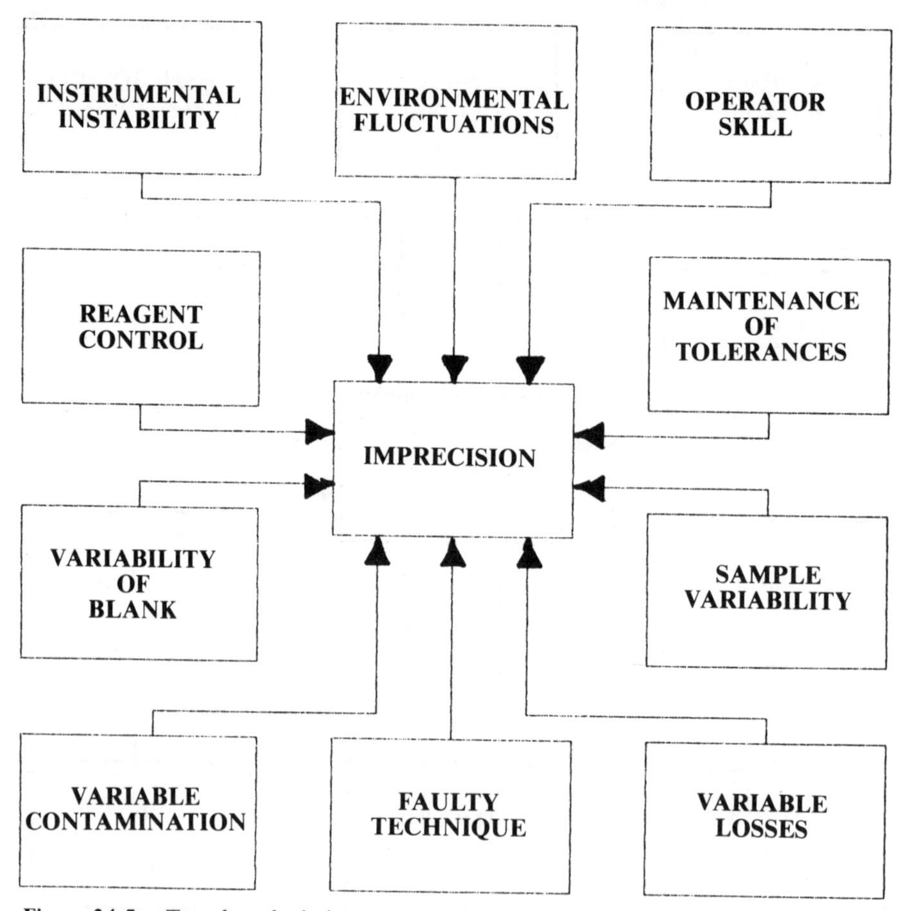

Figure 24.5. Template depicting sources of measurement imprecision.

can lead to retention of only those values that lie close to the majority. The author has known analysts who consistently practiced the "best two out of three" rule for reporting data. That is to say, three measurements were taken, and the two closest together were accepted. It has been shown that such a practice, rather than improving data, actually increases the variance of data sets.

Analysts are urged to retain all data unless there is clear indication that the system is not in control or there is an assignable cause for faulty data. In the absence of the above, statistically supported identification of outlying data is also justification for rejection. In any case, good laboratory practice requires that all data be recorded and that any data rejected should be clearly identified with the reason clearly stated.

An apparent exception to the above is the fact that some instrumentation requires a warmup period before it stabilizes and attains statistical control.

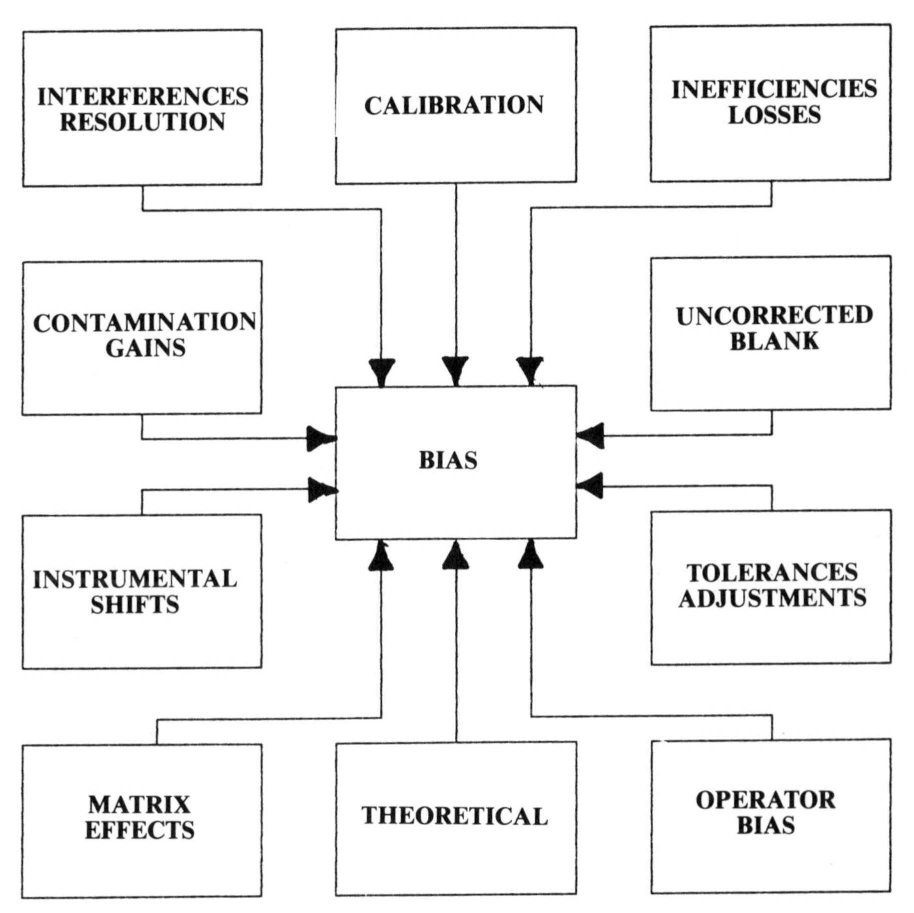

Figure 24.6. Template depicting sources of measurement bias.

The procedure to be followed in such cases should be described in the SOP and precisely followed. It is considered good practice in these cases to record such preliminary data with proper notation.

CHAPTER 25

Laboratory Evaluation

Unfortunately, not all laboratories are peers; hence, an evaluation of the capability of a laboratory to provide a desired service is an important consideration. Any evaluation requires a set of standards that define minimum acceptable performance and judgement as to whether the standards have been met. The evaluation may be performed by an individual requiring an individual service, or it may be required by some organization or agency as a basis for acceptance of services and/or data. In this case, the evaluation may result in accreditation or certification of capability by a third-party organization. Laboratories may voluntarily seek accreditation or certification to provide assurance to prospective clients of their capability.

Several organizations have given considerable attention to the development of criteria by which a laboratory may be evaluated. These range from generic [1,17] to specific for laboratories engaged in a defined class of measurements [8,10,11,19]. In many cases, even the above criteria are not considered to be sufficiently specific; hence, criteria are developed for a specific measurement situation. The criteria cited above have another function—namely, to provide a basis for self-evaluation by an individual laboratory.

GENERIC GUIDELINES FOR EVALUATION

The generic guidelines contained in ASTM Standard Practice E 548, summarized below, are followed by most accreditation systems and are recommended for laboratories to use for self-appraisal, as will be discussed later.

Organization: A well-organized and well-managed laboratory is necessary though not sufficient for data reliability. The duties and responsibilities of management and staff need to be well defined by position descriptions and organizational charts. Adequate supervision must be provided. A system to constantly inspect and evaluate outputs should be in place. Internal audits should be conducted periodically to monitor quality assurance and measurement practices. A self-appraisal system should be in effect whether or not external appraisal is utilized.

A well-managed laboratory should have available a list of the services it is competent to provide, together with supporting evidence to back up its claims. This may consist of the results of measurement of known reference materials, collaborative test data, and control charts, for example. Documented limits of uncertainty that can be expected should be available.

Staff: The staff should have technical competence and experience, commensurate with the difficulty and complexity of the services offered. Their qualifications should be a matter of record. Both formal and informal training should be provided to attain and maintain technical competence.

Equipment: Both general purpose and specialized equipment should be adequate in kind, quantity, and quality to satisfy measurement requirements. This must be properly housed and maintained and in a ready or standby condition for use in the services offered.

Calibration and reference standards: Calibration standards and reference materials that are adequate for the accuracy requirements of the services offered should be available. These should be properly stored so that their integrity will be maintained.

Test methods and procedures: The methods and procedures to be used in all measurement services should be listed and on hand as SOPs. Personnel, on questioning, should have full understanding and familiarity with them. Methods routinely used should be supported by control charts or other evidence of proficiency in their use.

Facilities: Physical facilities should be adequate with respect to space, environmental control, and maintenance. Any specialized requirements, such as clean rooms, should be available. Good housekeeping practices should be evident.

Test items: Test items should be properly handled and stored. A system of sample management should be in place. A chain of custody should be in operation, consistent with the requirements of the most critical samples handled.

Records: Records should be maintained adequately in kind and detail. A records management system should be in place. The ease with which information can be retrieved provides important insight into the adequacy of the records management system.

Test reports: These should be adequate in content and clarity. They should be keyed in with the test results and to all supporting evidence. Limits of uncertainty should be assigned for each test result and included in the report, numerically or by suitable codes or statements. A system for review and release of reports should be operational.

Quality assurance: A formal quality assurance program should be in operation which addresses and monitors many of the above requirements.

ACCREDITATION

There are basically two systems of accreditation of laboratories. One may be called "product focus" and the other "discipline focus." The former accredits or certifies capability to do rather specific measurements while the latter is more general. It is virtually impossible and would be meaningless to accredit or certify on an unconditional basis. However, a well-qualified laboratory should have reasonably broad capabilities. Herein lies the philosophical difference between the two systems. The former is based on the premise that only specific certification is meaningful, while the latter considers that the ability to measure in one area can infer ability to do so in related areas. If "related" is narrowly defined, then the latter system approaches the former.

The product focus system uses relatively narrow criteria for judgement. Thus, competence in the use of specific methods and the availability of specific equipment may be required. Laboratory facilities meeting specific requirements may be another criterion. Proficiency testing is a common but not necessarily a unique feature of product-focussed accreditation processes. It is rather easy, conceptually, to specify the kind of test samples for this purpose in this case because the products tested are narrowly defined. Generic test samples can be used for proficiency testing in discipline-focussed accreditation, but good judgment must be involved in their selection or else the testing is of limited value.

The parallelisms of the appraisal processes used by each system are evident on inspection of Figure 25.1. The appraisal process often begins with a self-appraisal by the laboratory which answers a questionnaire regarding various matters, including compliance with some mandatory requirements. If certain of these are not met, certification is not possible and the process stops. This may be followed by an onsite inspection to verify certain details and/or to obtain additional information. Certain judgements on operational matters can best be made on the basis of inspections and discussions with staff and management. Proficiency testing is done only after a laboratory passes essentially all other requirements. An appraisal based on definite criteria should result in an accreditation, or specific deficiencies should be identified whereby correction of them would lead to accreditation. Accreditation is ordinarily for a prescribed period of time or until some problem that would merit loss of accreditation occurs. Each system has specific requirements for renewal of accreditation. There are literally scores of accreditation systems in the United States, some of which differ only in small detail from one another, and yet reciprocity of certification across agencies and/or organizations is rare. Performance testing on test samples is often a requirement, but there is little if any cooperation between accreditors in this respect. The need of multiaccreditation for the same kind of measurements and the requirement for additional accreditation when a closely related test is offered are some of the problems that hinder widespread laboratory accreditation.

A further complication involves international accreditation. Organizations in the international market may need certification that a product meets specifications, but data provided by a competent laboratory in one country may not be acceptable elsewhere. Variable quality assurance practices of various laboratories have been barriers to international acceptance of data, but differences in arbitrary requirements for accreditation have been unduly restrictive. International efforts to develop uniform criteria for evaluation may ease some of the above barriers even if national barriers exist.

SELF-EVALUATION

Self-evaluation is a hallmark of a well-managed laboratory. It has been said that a skeptical analytical chemist is the only one worthy of the name. The

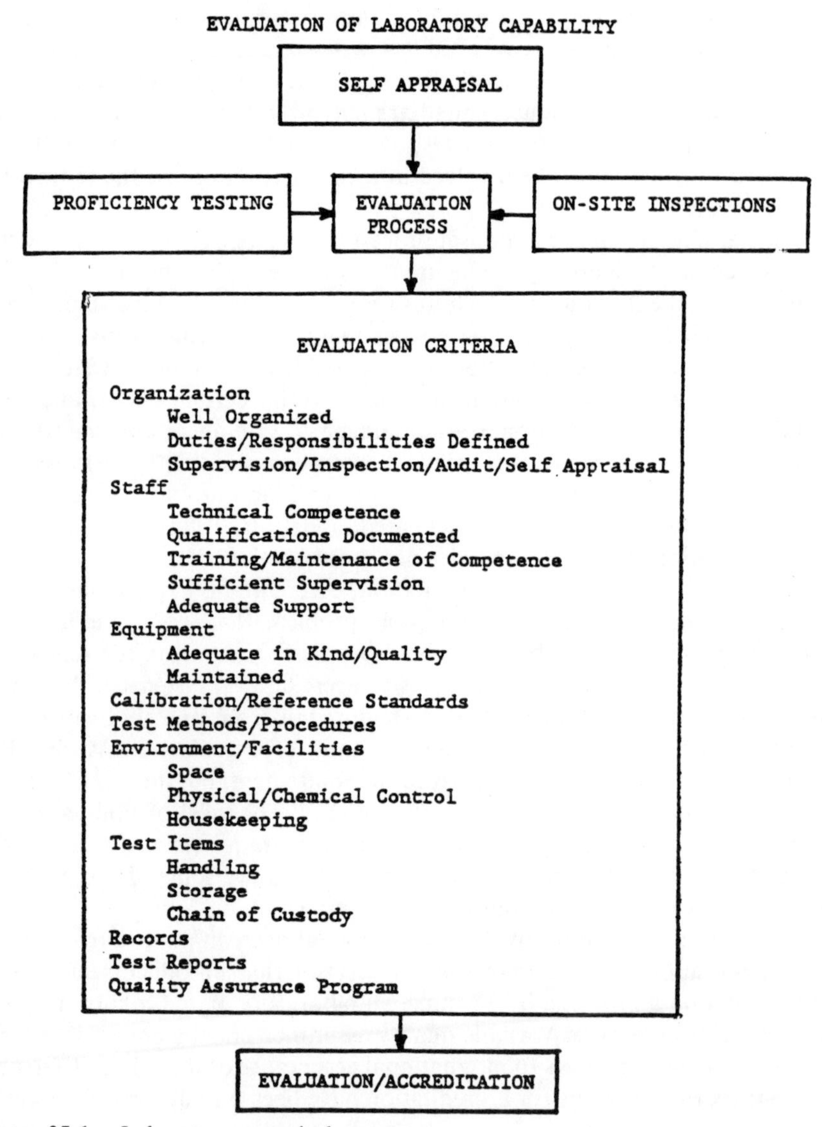

Figure 25.1. Laboratory appraisal process.

Table 25.1. Personal Quality Assurance Responsibilities (an Analyst's / Laboratory Code of Ethics)

1. Acquire a full understanding and develop peer technical expertise in every area of analytical chemistry in which professional services are offered.
2. Comprehend, to the extent possible, all problems for which analytical services are required: ascertain the validity of the approach selected; understand any limitations on the measurements and discuss them with clients as appropriate.
3. Use validated methodology exclusively.
4. Demonstrate statistical control of the measurement system before definitive measurements are made.
5. Calibrate, to the extent necessary and possible, and engage in intercalibration activities as appropriate to minimize chance of internal laboratory bias.
6. Utilize recognized Good Laboratory Practices (GLPs) and Good Measurement Practices (GMPs) throughout all aspects of sampling and measurement processes.
7. Utilize documented procedures (SOPs) and record all significant experimental details in such a way that the measurements could be reproduced by a competent analyst at a later date as necessary.
8. Provide or make available limits of uncertainty of all data reported, including that due to sample and to measurement, supported by statistical inference and/or professional judgement as pertinent; state clearly the basis for any interpretations provided of the measurement data.
9. Confirm the qualitative identification of all parameters measured and provide supporting evidence as necessary.
10. Retain all samples, data, and documentary evidence as necessary for a period of time commensurate with their importance.

same can be said for a laboratory. Accordingly, each laboratory should have a systematic plan for critical review and appraisal of its program and outputs. If quality assurance is mandatory, or certification is a requirement, a laboratory is well advised to audit its practices more critically than they will be examined by others. Even if there are no external requirements to be met, a laboratory should periodically review its compliance with generic standards. The generic guidelines referred to and outlined above provide an excellent basis for self-appraisal.

Regardless of whether certification is or is not required, a laboratory offering services has professional obligations that it should meet. A laboratory that subscribes to and fully conforms to the Quality Assurance Code of Ethics presented in Table 25.1 will go a long way toward meeting its professional obligations. Laboratories that have implemented a credible quality assurance program and have an ongoing internal auditing program as already described (see Chapter 19) to evaluate compliance and performance have the capacity to produce technically sound and legally defensible data.

CHAPTER 26

The Quality Assurance Program

Formally or informally, quality assurance is a necessary part of the production of data. Hardly anyone will quarrel with the concept that measurements should be made with appropriate care and that limits of uncertainty are necessary for their logical use. Only the question of how the goals are to be accomplished is debatable.

APPROACHES TO QUALITY ASSURANCE

There is a broad spectrum of approaches that can be followed to achieve quality assurance, the extremes of which are indicated in Table 26.1. At one end is the craftsman-artisan approach which depends on the use of highly motivated personnel operating in an atmosphere that encourages excellence. At the other end is the formal programmatic approach which defines all operational aspects of quality assurance.

Virtually no situation exists for which either approach is optimum. Instead, something between the two extremes must be considered. A high-level research effort approaches the craftsman-artisan situation but will be benefited by some aspects of a formal QA program. While highly formalized QA programs are applicable to most routine operations, these will be facilitated as craftsmanship is practiced. In fact, different operations in the same laboratory may vary in the kind of approach that would be most effective.

In deciding the details of a particular QA program, the costs and benefits of various approaches will need to be considered. Some of these are outlined in Table 26.2. As in many other activities, the costs may be easier to quantify than the benefits because the latter are often preventative. In fact, intangible benefits such as those listed in Table 26.3 may be the most important outcome of many quality assurance programs.

The size and diversity of an organization is an important determining factor in decisions related to formalization. Large operations concerned with narrowly defined programs of work are easily adapted to formal QA programs, and the development costs can be easily justified. Laboratories engaged essentially in small operations and/or those involved in diversified programs, on

Table 26.1. Approaches to Quality Assurance

CRAFTSMAN/ARTISAN APPROACH
Responsibility—Craftsman
Effectiveness
depends on craftsman's knowledge, skill, dedication
Requirements
highly skilled/dedicated craftsman
atmosphere which encourages excellence
Outstanding Characteristics
accuracy, redundancy
Works Best For
complex investigations
Controls
peer review
FORMAL QUALITY ASSURANCE PROGRAM
Responsibilities—Management
Effectiveness
depends on defined protocols, trained operators, strict compliance
Requirements
infallible protocols, competent staff
Outstanding Characteristics
high precision, efficiency
Works Best For
routine, recurring, well-defined problems
Controls
quality assurance program/office

first consideration, may conclude that a formal quality assurance program is virtually impossible.

For small laboratories or those with a varied workload, a critical study will often identify commonalities of a group of measurement operations for which generic QA practices may be applicable. Certainly, GLPs have wide applicability, and GMPs can be extended across different kinds of measurements made by the same technique. Considerations based on the common aspects of the type of matrix analyzed (e.g., water, sediments, metals) can help to unify operations for QA purposes. Identification of trouble spots (hot spot approach) related to diverse operations (e.g., contamination, reagents, housekeeping practices, and recordkeeping) can suggest system improvements and controls that can improve data quality. A laboratory should not look too narrowly at what it does. A comprehensive knowledge of the nature of the measurement processes used, an examination of how errors may be propagated, and facilitation in identifying assignable causes can aid in the development of credible QA programs for both large and small operations.

For complex operations, or those involving a number of participants or numerous repetitive measurements, a formal quality assurance program is

Table 26.2. Cost/Benefits of Quality Assurance

COSTS
DIRECT
Test Materials
Standards
Quality Assurance Equipment—Test Instruments
Analysis of Quality Control/Quality Assurance Samples
Time of Personnel
Time of Supervision
Quality Assurance Official
Committee Work
Round Robin Costs
Travel/Attendance at Meetings
INDIRECT
Training
Extra Cost for Quality People
Extra Quality Equipment
Extra Quality Supplies
Relaxed Work Schedules
BENEFITS
More Efficient Outputs
Fewer Replicates for Same Reliability
Fewer Do-Overs
Greater Confidence Of:
Staff
Laboratory
Customers

Table 26.3. Intangible Benefits of a QA Program

Promote External Image	Eliminate Unnecessary Redundancies
Improve Internal Image	Promote Continuity of Effort
Promote Client Confidence	Provide for Retention of Vital Records
Add Credence to Results	Set Forth Goals and Objectives
Prevent Hasty Disclosures	Provide Guidance to Staff
Minimize Indecision	Provide Basis for Training

virtually a necessity, and the time and effort devoted to its development should be more than compensated by cost-effective and highly reliable analytical data. Indeed, the overall cost of operation should not exceed 10% of the total operational effort, and "do-overs" and "lost" data should be minimized. On this basis, in production operations, it has been said that "quality is free" [40].

In research investigations, an informal quality assurance program is often considered adequate. However, the principles of quality control and quality assessment cannot be ignored in such work, and their implementation, even in an informal way, could constitute a major portion of the experimental effort. Accordingly, research organizations can formalize appropriate recurring parts of their programs and substantially reduce the overall quality assurance effort in many cases.

QUALITY ASSURANCE POLICY

The adoption and enunciation of a quality assurance policy by management is a prerequisite for an effective quality assurance program. This will reflect the degree of commitment by management to quality outputs and the resources that will be made available for this purpose. Based on the policy, a program can be developed and implemented. The policy should define the respective responsibilities of the various organizational levels in implementing the program.

PLANNING FOR QUALITY ASSURANCE

Some degree of quality assurance is commonly practiced in most laboratories. Often, this is most of what is needed if the past measurement process has been well planned. Ordinarily, there is little opposition by laboratory personnel to doing a good job. However, there is a natural aversion to regulations and requirements, even when they merely describe what is ordinarily done. When regulations include trivia or incredible requirements, they not only discredit the important ones, but also engender contempt for the entire program. Accordingly, the quality assurance program must be both realistic and perceived as such.

The building blocks for a quality assurance program have been described earlier [132, 137] and are illustrated in Figure 26.1. The specificity increases in descending order. The QA policy of a laboratory reflects its basic measurement philosophy and the goals and objectives it desires to meet. It will reflect the type of services it strives to render and the desires and needs of the clientele it serves. As already discussed, the GLPs and GMPs describe in rather specific terms the ways certain operations will be conducted. SOPs define how specific measurements will be made, and the PSPs encompass all of the above for specific measurement activities.

A laboratory's quality assurance program should be internally motivated by the general desire to produce high-quality outputs, but it is externally moti-

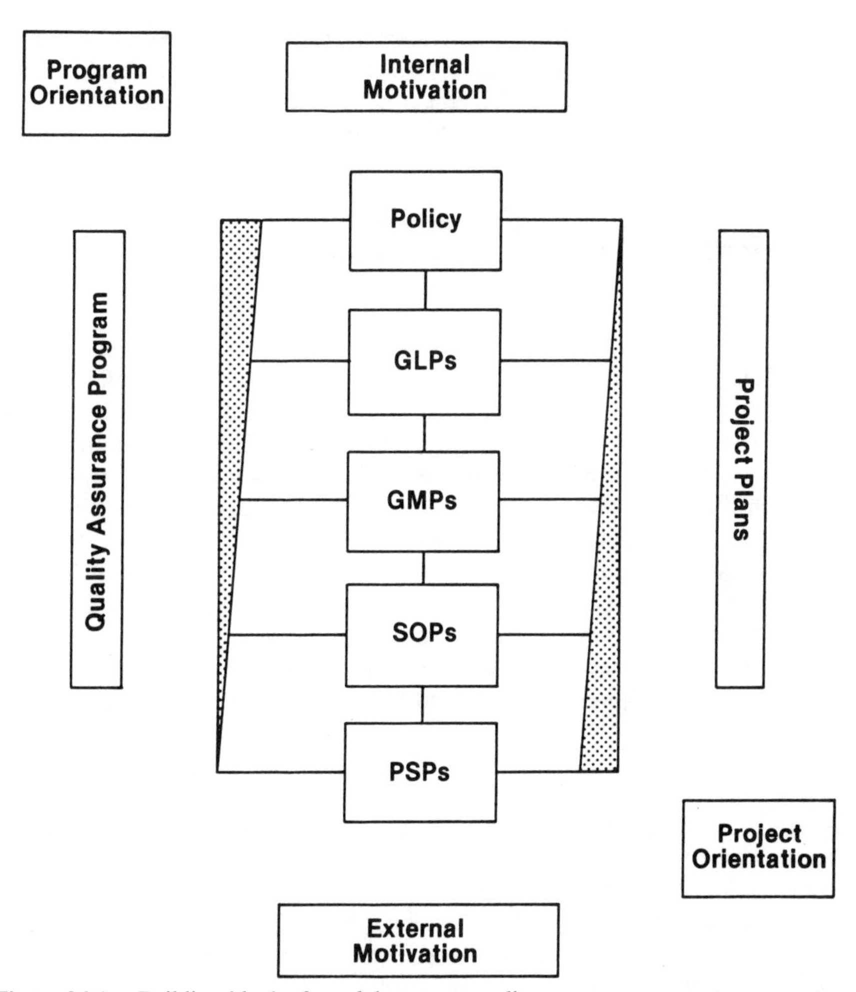

Figure 26.1. Building blocks for a laboratory quality assurance program.

vated as well by specific requirements for a specific project. The former is thus program oriented and usually requires the development of a quality assurance manual. The latter is problem oriented and requires the development of project plans. The density of the tie-lines in the figure indicates the degree of relatedness of the central elements to the ongoing program (left side) or to a specific project (right side).

A laboratory's quality assurance program, as delineated in its quality assurance manual, should be sufficiently comprehensive to apply to most of the measurement services which it can provide. Also, it may include PSPs for its recurring measurement activities. Ordinarily, PSPs will need to be developed for a specific project of a type not previously encountered and especially for monitoring programs. Even when PSPs exist, they may need to be reexamined for their adequacy at each time of reuse and revised as necessary. When a laboratory has a well-developed quality assurance program, PSP development can consist of the selection of elements appropriate to the project of concern with only minor modifications required to adapt them to the specific program.

DEVELOPMENT OF A QUALITY ASSURANCE PROGRAM

The development of a quality assurance program should involve all levels of laboratory personnel. The management must make the decision to develop a formal program and establish the policy that will be followed. In doing so, it should commit appropriate resources for both its development and implementation. Ordinarily, management designates a leader to developg the details of the program. Management will also need to approve the program at various stages of its development and establish a mechanism for its oversight and implementation.

The leader may or may not chair a committee, depending on the size of the organization. In any case, the leader must get the cooperation and involvement of the staff in developing the details and will follow appropriate procedures to obtain consensus approval.

The staff must provide technical advice and guidance, particularly in the development of the GLPs, GMPs, and SOPs. The staff may actually write appropriate parts of these documents or at least should review what is written. Consensus approval by the staff is a necessity for general acceptability and for the preclusion of trivial and unnecessary requirements. Appropriately constituted Quality Circles provide an excellent mechanism for developing the details of the program. Interdisciplinary groups can address the GLPs while disciplinary groups are best suited for developing GMPs. SOPs are best developed on a case-by-case basis.

Staff involvement will be found educational for them, and this may be one of the most important features of the development of a formal quality assurance program. The development process will require a careful consideration and comparison of the measurement process as it is practiced internally and in state-of-the-art laboratories. Parallelisms and contrasts between the practices of various groups and measurement techniques within a laboratory can be identified, and the opportunity is provided to elevate all operations to the highest common factor. When the quality assurance program is assembled into the form of a manual, it can provide future guidance as well as documentation of present activities.

The process of development of a quality assurance program need not be complicated. The developers should identify the goals to be accomplished and the motivation behind its development. This will include an examination of the internal policy and any QA requirements of external programs with which there is appreciable interaction.

The next step is to identify and document the quality assurance procedures already implemented, partially or fully, in the ongoing program. These should be compared with any requirements and also with the experience of others. Generic guidance is provided in earlier chapters of this book and in several of the articles in the bibliography (Appendix E) [1, 5, 6, 8, 10, 11, 17].

Obtaining consensus approval is the next step. This will include concurrence by the internal participants of the program, management, and any external group or organization requiring a formal QA program, in the order mentioned.

In all of the above, it may not be feasible to cover all of the activities of a laboratory in the first iteration. In fact, if too much is attempted, it could lead to frustration and be counterproductive. A laboratory may need to set priorities based on external demands, the criticality of operations, or what can be readily formalized from a present program. Implementation of the quality assurance program is as important as its development. All levels of the laboratory staff are involved, and the responsibilities of each should be clearly defined in the quality assurance program document. Management may designate a quality assurance officer or coordinator to oversee the program. In some situations, this may be mandatory. The responsibilities of such an officer, outlined in Table 26.4, may range from periodic reviewing of the program and conducting systems audits to providing test materials and conducting internal performance audits on a routine and continuing basis. The review and approval of the quality assurance aspects of proposed work and development of the PSPs for such may also be a responsibility of this person or office. The quality assurance office should be independent of daily operations and report directly to management so that unbiased appraisals of operations can result.

Middle management (the supervisory staff) has the responsibility for direct implementation of the quality assurance program. Work of subordinates is reviewed for its technical excellence and conformance with quality assurance requirements. Middle management provides training related to both aspects and makes decisions on corrective actions, including the identification of assignable causes for measurement-process deficiencies.

The individual members of the technical staff have key responsibilities for the quality of the data output and for all related laboratory operations. The staff must be adequately trained to carry out both its technical and quality assurance responsibilities. Staff members often have the first opportunity to detect and correct malfunctions of the measurement system and must be trained to do this. Both the technical and supervisory staff must be able to distinguish between normal random fluctuations and abnormal deviations. Unnecessary overcorrections can often increase the variance of the measurement process and should be avoided. The QA program should provide definite instructions in this regard.

The quality assurance program should be strictly enforced, yet there should

Table 26.4. Duties of a Quality Assurance Officer

Basic Function: The quality assurance officer is responsible for the conduct of the quality assurance program and for taking or recommending corrective actions as required.

Responsibilities and Authority

1. Develops and carries out quality control programs, including statistical procedures and techniques which will meet desired quality standards at minimum cost.
2. Monitors quality assurance activities to determine conformance with policy and procedures and with sound practices; conducts system audits and makes appropriate recommendations for corrective actions and improvements as may be necessary.
3. Seeks and evaluates new ideas and current developments in the field of quality assurance and recommends means for their application whenever advisable.
4. Advises management in reviewing technology, methods, equipment, and facilities with respect to their quality assurance aspects.
5. Coordinates schedules for measurement systems functional checks, calibrations, and other checking procedures.
6. Develops system for control sample measurement; advises in the selection of suitable samples and/or prepares samples as delegated by management.
7. Evaluates data quality and maintains records and/or control charts, calibration records, and other pertinent performance information.
8. Coordinates and/or conducts quality-problem investigations.
9. Advises and trains staff in matters of quality assurance.
10. Reviews work proposals for quality assurance aspects.
11. Serves as organization's quality assurance representative for mandated QA programs.

be flexibility to deviate from it as necessary. If deviations are necessary too frequently, the program may be defective and require amendment. Deviation may be required because of both underspecification and overspecification of the quality assurance program, or by changes in the measurement program that were not anticipated or are not covered by the program as originally developed. For such reasons, the program should be continually reviewed and updated as required. While flexibility should be provided, the objective should be to develop a quality assurance document that can be followed with confidence and for which deviations would be exceptional and infrequent.

QUALITY ASSURANCE MANUAL

The quality assurance program and practices of a laboratory should be documented. Ordinarily this will take the form of a manual which will include:

- the QA program document
    - GLPs
    - GMPs
    - SOPs
- implementation directives

The quality assurance manual may be viewed as an instruction kit that the laboratory itself has developed to guide its laboratory operations in the production of quality data. It can provide internal guidance on policy and procedures and serve as a training manual for new employees. It should satisfy most external requirements as well, so that its existence will minimize effort wherever PSPs need to be developed.

The quality assurance manual should be comprehensive but concise, and descriptive but brief. Because some parts will be general and others specific, it will need to be well organized and indexed. QA manuals may be more useful when they are entered into computers. Careful indexing can facilitate computer retrieval of information pertinent to a specific measurement problem and thus enhance its usefulness.

The various documents needed to formalize a quality assurance program should be prepared with care. The general outlines for documents, contained in Appendix B, may be helpful when preparing QA program documentation.

FEDERALLY MANDATED QUALITY ASSURANCE PROGRAMS

The general aspects of the major federally mandated quality assurance program are described briefly in the following sections. Laboratories that must comply with them should consult the Federal Register citations for details. The requirements contained therein are fundamentally sound and in complete harmony with the philosophy advocated in this book. Laboratories offering measurement services in related areas, whether or not directly affected, will find these regulations instructive for developing or strengthening their analytical programs.

Public Health Service

Clinical Laboratories

The Clinical Laboratories Improvement Act of 1967 [38] contains requirements for quality assurance as part of the general licensing requirements for such laboratories. It is applicable to laboratories engaged in examination of specimens solicited or accepted in interstate commerce for the purpose of providing for the diagnosis, treatment, or assessment of the health of man. It does not apply to small operations (less than 100 specimens per calendar year) or data used by a physician in his own practice, or to insurance examinations. It does not apply to an accredited laboratory, provided that the Secretary finds that the standards applied are equal to or are more stringent than those of the Act. The requirements, which may be classified as general- and discipline-specific, respectively, are summarized below.

General. Quality controls imposed and practiced must provide for and assure:

- Inspection and preventative maintenance of equipment and instruments; periodic inspection and testing for proper operation
- Validation of methods
- Remedial actions taken in response to detected defects
- Monitoring of temperature-controlled spaces and equipment
- Evaluation of analytical measuring devices
- Labeling of all reagents and solutions
- Banning the use of substandard and deteriorated materials
- Availability of a laboratory procedural manual; exclusive use of documented methods
- Manuals located in the immediate bench area
- Manuals current and designed to reflect an annual supervisory review
- Manuals and methods specify calibration procedures
- Manuals and methods include written approval of all changes
- A list of all analytical methods applied on file and available for inspection of concerned parties

Chemistry. The following requirements, in addition to the general requirements, relate to all chemical measurements of the laboratory:

- Records document routine precision of each method used
- At least one standard and one reference sample included in each run
- Control limits recorded
- A written course of action followed when controls are outside of acceptable limits
- Minimum detection limits available for toxicology
- Qualitative chemical analysis checked daily with reference samples
- Reports of quantitative analysis include the units of concentration or activity
- When kits are used, each shall be tested when opened and at least once each week of use

Others: There are specific requirements for measurements in the fields of bacteriology/mycology; parasitology; virology; serology; immunohematology; hematology; cytology; histology/oral pathology; radioassay.
Records: The following requirements pertain to all fields.

- Records of observations shall be made concurrently with the performance of each step in the examination of specimens.
- Records shall reflect the actual results of all control procedures.
- Records shall be retained for two years.
- Personnel records shall be maintained on a current basis.
- Daily accession records shall be maintained, containing: laboratory number, name of patient, name of physician, date specimen was collected, date specimen was received in the laboratory, the condition of unsatisfactory specimens, type of test performed, and results of test and date of completion.
- The name and address of the laboratory actually performing the test shall be indicated in the report.

Food and Drug Administration

Nonclinical Laboratories

The Food and Drug Administration has promulgated regulations known as good laboratory practices [36], applicable to laboratories that support applications for research or marketing permits for products regulated by the FDA, including:

- food and food additives
- animal food additives
- human and animal drugs
- medical devices for human use
- biological products
- electronic products

The regulations are also applicable to consulting laboratories, contractors, and grantees related to the above activities. Many of the requirements are similar to those for clinical laboratories. Those specific to this Act are outlined below. **Organization and Management**. Requirements pertaining to laboratory organization and management include the following:

- Inspections for compliance must be permitted.
- All personnel must have education, training and experience adequate for their duties; job descriptions, resumes, and training records must be on file for all personnel; sufficient staff must be available to conduct studies; clothing must be appropriate for duties and for preventing contamination.
- Management designates a Study Director, establishes a Quality Assurance Unit (QAU), is responsible for validating results, assures availability of staff, and assures that deficiencies are corrected.
- Study Director must have appropriate training and experience, has overall responsibility for conduction of study, assures that protocols are followed, and is solely responsible for authorizing changes in protocols.

- Quality Assurance Unit is mandatory and separate from study management; is responsible for assuring management that regulations are followed; maintains master schedule chart; maintains copies of all protocols that must be followed; periodically inspects each phase of a study for compliance; submits periodic reports to management, noting problems and corrective actions needed; determines that no unauthorized deviations from protocols have occurred; reviews final report; prepares statement for final report regarding interim inspections; and maintains records of above for use of FDA on request.
- Facilities must be suitable for measurements and handling of test items; administrative and personnel facilities must be adequate, including sanitation facilities as appropriate.
- A written protocol is required for each study that defines all aspects of how it is to be conducted and the records that are to be kept; protocol must be approved by Study Director and sponsor as appropriate; study must be conducted according to the protocol and monitored for conformance; and any changes must be explainable.
- Final report must contain full information, specified in regulation; raw data, documentation, protocols, specimens, and final report must be retained; and various retention times are specified, depending on the nature of the records.
- Disqualification will result in nonacceptance of data; various grounds for disqualification are specified; and reinstatement possible.

Environmental Protection Agency

General Requirements

It is agency policy that all work performed by the Agency in its own laboratories or under its sponsorship, directly or indirectly, shall be conducted according to a Quality Assurance Project Plan [125]. The plan must be developed and approved in advance of the work and address the following sixteen items:

(1) Title page with provision for approval signatures
(2) Table of contents
(3) Project description
(4) Project organization and responsibilities
(5) Quality assurance objectives for measurement data in terms of precision, accuracy, completeness, representativeness, and comparability
(6) Sampling procedures
(7) Sample custody
(8) Calibration procedures and frequency
(9) Analytical procedures
(10) Data reduction, validation, and reporting

(11) Internal quality control checks and frequency
(12) Performance and system audits and frequency
(13) Preventative maintenance procedures and schedules
(14) Specific routine procedures to be used to assess data precision, accuracy, and completeness of specific measurement parameters involved (mandatory in every plan)
(15) Corrective actions
(16) Quality assurance reports and management

Toxic Substances Control

In addition to the above, specific regulations have been established, describing good laboratory practices for conducting chemical studies of health effects, environmental effects, and chemical fate testing [37]. These are essentially identical with the FDA good laboratory practices outlined above.

NONFEDERAL MANDATED QUALITY ASSURANCE PROGRAMS

Laboratories are becoming increasingly aware that they must comply with the quality assurance requirements of others in order to qualify to supply services. This can be direct in the case of data, or indirect when they evaluate the properties of products that must meet designated quality standards. While a laboratory with a good quality assurance program should meet almost any reasonable general requirements, there could be specific ones that would necessitate special procedures. For example, specific methodology may be necessary in certain cases, and specific control charts may need to be maintained. A laboratory may need to demonstrate peer performance by successful measurement of test samples as a prerequisite for award of a contract. Also, there may be specific requirements for chain of custody of samples and special requirements for recordkeeping and retention. A laboratory should be alert for such requirements and make proper adjustments to its program if it wants to participate in such activities.

The Deming philosophy of "know your subcontractors" [43] is fast becoming a practice in industry. Manufacturers are quality rating their suppliers and buying on a quality rather than a price basis. The inspection function for compliance with specifications is being passed down to suppliers along with the requirement that this be supported by statistical control of the production process. When measurement is a part of the production process, as is almost always the case, the above requirements apply as well.

No matter what the situation—production of materials, services, or measurement data—quality assurance practices will improve the odds for obtaining reliable products and services. With all other considerations equal, an effective quality assurance program can be decisive in the selection of a supplier. Vendors that are required to comply with quality assurance provisions generally perform better than those who are not so governed.

QUALITY ASSURANCE FOR MEASUREMENT PROGRAMS

Measurement programs in which data are collected over a period of time for monitoring or historical purposes must ensure that compatibility is achieved. The need for a credible quality assurance program to ensure this is obvious [138]. All producers of such data must demonstrate competence before their data can be accepted. When several participants are involved, peer performance of all contributors must be achieved. Minimum performance standards must always be maintained and demonstrated on a continuing basis and should be the basis for validation and acceptance of data, and indeed payment for it. In large programs, adherence to the above requirements may be best monitored and the validation of data facilitated by the use of a reference laboratory. One of the major functions of such a laboratory is to provide reference materials and various control samples to ensure the continuing intercalibration of the participating laboratories.

RELATING PERSONNEL TO QUALITY ASSURANCE

The goal of introducing a quality assurance program into an organization is acceptance rather than resignation. In fact, the motivations for improving and/or formalizing quality assurance activities range from professional pride, the risk of being wrong, the consequences of adverse actions, and even the threat of malpractice accusations. Federal regulations are sometimes undebatable reasons for adopting a formal QA program [36, 37, 38, 47]. The international arena, governed by the General Agreement on Tariffs and Trade (GATT) and other international agreements, may mandate a formal QA program.

When regulations are the motivation, the technical staff should be informed of the requirements so that realistic acceptance can result. Where voluntary considerations are involved, the staff may need to be prepared in a different way. Their involvement in the development of GLPs and GMPs instills a spirit of ownership as well as the development of realistic practices.

Quality assurance activities should be perceived as useful and not as punitive measures. They can and should enhance the quality of outputs and provide the basis for their evaluation. A quality assurance program developed by consensus should be credible and perceived as such. No matter how adequate at the time of its development and adoption, its use may identify the need for additions or amendments, and prompt actions of this nature should be taken as necessary.

APPENDIX A

Terminology Used in Quality Assurance[1]

Absolute method—A method in which characterization is based entirely on physically (absolute) defined standards.

Accreditation—A formal process by which a laboratory is evaluated, with respect to established criteria, for its competence to perform a specified kind(s) of measurement. Also, the decision based upon such a process. When a certificate is issued, the process is often called certification.

Accuracy—The degree of agreement of a measured value with the true or expected value of the quantity of concern.

Aliquant—A divisor that does not divide a sample into a number of equal parts without leaving a remainder; a sample resulting from such a divisor.

Aliquot—A divisor that divides a sample into a number of equal parts, leaving no remainder; a sample resulting from such a divisor.

Analyte—The specific component measured in a chemical analysis; also called analate.

Assignable cause—A cause believed to be responsible for an identifiable change of precision or accuracy of a measurement process.

Blank—The measured value obtained when a specified component of a sample is not present during the measurement. In such a case, the measured value/signal for the component is believed to be due to artifacts; hence, it should be deducted from a measured value to give a net value due to the component contained in a sample. The blank measurement must be made so that the correction process is valid.

[1]For other definitions, and particularly those related to industrial quality assurance, see: "Quality Systems Terminology," ANSI/ASQC Standard A3-1978, American National Standards Institute, 1430 Broadway, New York, NY 10018.

Blind sample—A sample submitted for analysis whose composition is known to the submitter but unknown to the analyst. A blind sample is one way to test proficiency of a measurement process.

Bias—A systematic error inherent in a method or caused by some artifact or idiosyncrasy of the measurement system. Temperature effects and extraction inefficiencies are examples of this first kind of bias. Blanks, contamination, mechanical losses and calibration errors are examples of the latter kinds. Bias may be both positive and negative, and several kinds can exist concurrently, so net bias is all that can be evaluated except under special conditions.

Bulk sampling—Sampling of a material that does not consist of discrete, identifiable, constant units, but rather of arbitrary, irregular units.

Calibrant—A substance used to calibrate or to establish the analytical response of a measurement system.

Calibration—Comparison of a measurement standard or instrument with another standard or instrument to report or eliminate by adjustment any variation (deviation) in the accuracy of the item being compared.

Cause-effect diagram—A graphical representation of the causes that can produce a specified kind of error in measurement. A popular one is the so-called fish bone diagram, first described by Ishikawa, given this name because of its suggestive shape.

Certification—See Accreditation.

Central line—The long-term expected value of a variable displayed on a control chart.

Certified reference material (CRM)—A reference material, one or more of whose property values are certified by a technically valid procedure, accompanied by or traceable to a certificate or other documentation which is issued by a certifying body.

Certified value—The value that appears in a certificate as the best estimate of the value for a property of a reference material.

Chance cause—A cause for variability of a measurement process that occurs unpredictably, for unknown reasons, and believed to happen by chance.

Check standard—In physical calibration, an artifact measured periodically, the results of which typically are plotted on a control chart to evaluate the measurement process.

Coefficient of variation—The standard deviation divided by the value of the parameter measured.

Common cause—A cause of variability of a measurement process, inherent in and common to the process itself, as contrasted to a special cause.

Comparative method—A method which is based on the intercomparison of the sample with a chemical standard.

Confidence interval—That range of values, calculated from an estimate of the mean and the standard deviation, which is expected to include the population mean with a stated level of confidence. Confidence intervals in the same context may also be calculated for standard deviations, lines, slopes, and points.

Control limit—The limits shown on a control chart beyond which it is highly improbable that a point could lie while the system remains in a state of statistical control.

Control chart—A graphical plot of test results with respect to time or sequence of measurement, together with limits within which they are expected to lie when the system is in a state of statistical control.

Control sample—A material of known composition that is analyzed concurrently with test samples to evaluate a measurement process (see also Check standard).

Composite sample—A sample composed of two or more increments selected to represent a population of interest.

Cross sensitivity—A quantitative measure of the response obtained for an undesired constituent (interferant) as compared to that for a constituent of interest.

Detection limit—The smallest concentration/amount of some component of interest that can be measured by a single measurement with a stated level of confidence.

Double blind—A sample known by the submitter but submitted to an analyst in such a way that neither its composition nor its identification as a check sample are known to the latter.

Duplicate measurement—A second measurement made on the same (or identical) sample of material to assist in the evaluation of measurement variance.

Duplicate sample—A second sample randomly selected from a population of interest to assist in the evaluation of sample variance. (See also Split sample.)

Education—Disciplining the mind through instruction or study. Education is general and prepares the mind to react to a variety of situations.

Error—Difference between the true or expected value and the measured value of a quantity or parameter.

Figure of merit—A performance characteristic of a method believed to be useful when deciding its applicability for a specific measurement situation. Typical figures of merit include: selectivity, sensitivity, detection limit, precision, and bias.

Good laboratory practice (GLP)—An acceptable way to perform some basic operation or activity in a laboratory that is known or believed to influence the quality of its outputs. GLPs ordinarily are essentially independent of the measurement techniques used.

Good measurement practice (GMP)—An acceptable way to perform some operation associated with a specific measurement technique and known or believed to influence the quality of the measurement.

Gross sample (also called bulk sample, lot sample)—One or more increments of material taken from a larger quantity (lot) of material for assay or record purposes.

Homogeneity—The degree to which a property or substance is randomly distributed throughout a material. Homogeneity depends on the size of the subsample under consideration. Thus a mixture of two minerals may be nonhomogeneous at the molecular or atomic level but homogeneous at the particulate level.

Increment—An individual portion of material collected by a single operation of a sampling device from parts of a lot separated in time or space. Increments may be either tested individually or combined (composited) and tested as a unit.

Individuals—Conceivable constituent parts of a population.

Informational value—Value of a property, not certified but provided because it is believed to be reliable and to provide information important to the certified material.

Intercalibration—The process, procedures, and activities used to ensure that the several laboratories engaged in a monitoring program can produce compatible data. When compatible data outputs are achieved and this situation is maintained, the laboratories can be said to be intercalibrated.

Laboratory sample—A sample intended for testing or analysis prepared from a gross sample or otherwise obtained. The laboratory sample must retain the composition of the gross sample. Often, reduction in particle size is necessary in the course of reducing the quantity.

Limiting mean—The value approached by the average as the number of measurements made by a stable measurement process increases indefinitely.

Limit of linearity (LOL)—The upper limit of concentration or amount of substance for which incremental additions produce constant increments of response.

Limit of quantitation (LOQ)—The lower limit of concentration or amount of substance that must be present before a method is considered to provide quantitative results. By convention, $LOQ = 10s_o$, where s_o is the estimate of the standard deviation at the lowest level of measurement.

Lot—A quantity of bulk material of similar composition whose properties are under study.

Method—An assemblage of measurement techniques and the order in which they are used.

Outlier—A value which appears to deviate markedly from that for other members of the sample in which it occurs.

Pareto analysis—A statistical approach to ranking assignable causes according to the frequency of occurrence.

Performance audit—A process to evaluate the proficiency of an analyst or laboratory by evaluation of the results obtained on known test materials.

Population—A generic term denoting any finite or infinite collection of individual things, objects, or events; in the broadest concept, an aggregate determined by some property that distinguishes things that do and do not belong.

Precision—The degree of mutual agreement characteristic of independent measurements as the result of repeated application of the process under specified conditions. It is concerned with the closeness of results.

Primary standard—A substance or artifact, the value of which can be accepted (within specific limits) without question when used to establish the value of the same or related property of another material. Note that the primary standard for one user may have been a secondary standard of another.

Probability—The likelihood of the occurrence of any particular form of an event, estimated as the ratio of the number of ways or times that the event may occur in that form to the total number of ways that it could occur in any form.

Procedure—A set of systematic instructions for using a method of measurement or sampling or of the steps or operations associated with them.

Protocol—A procedure specified to be used when performing a measurement or related operation as a condition to obtain results that could be acceptable to the specifier.

Protocol for a specific purpose (PSP)—Detailed instructions for the performance of all aspects of a specific measurement program. This is sometimes called a project QA plan.

Quality—An estimation of acceptability or suitability for a given purpose of an object, item, or tangible or intangible thing.

Quality assessment—The overall system of activities whose purpose is to provide assurance that the quality control activities are done effectively. It involves a continuing evaluation of performance of the production system and the quality of the products produced.

Quality assurance—A system of activities whose purpose is to provide to the producer or user of a product or service the assurance that it meets defined standards of quality. It consists of two separate but related activities, quality control and quality assessment.

Quality circle—A small group of individuals with related interests that meets at regular intervals to consider problems or other matters related to the quality of outputs of a process and the correction of problems or the improvement of quality.

Quality control—The overall system of activities whose purpose is to control the quality of a product or service so that it meets the needs of users. The aim is to provide quality that is satisfactory, adequate, dependable, and economic.

Random sample—A sample selected from a population, using a randomization process.

Reduction—The process of preparing one or more subsamples from a sample.

Reference material (RM)—A material or substance, one or more properties of which are sufficiently well established to be used for the calibration of an apparatus, the assessment of a measurement method, or for the assignment of values to materials.

Reference method—A method which has been specified as capable, by virtue of recognized accuracy, of providing primary reference data.

Relative standard deviation—The coefficient of variation expressed as a percentage.

Replicate—A counterpart of another, usually referring to an analytical sample or a measurement. It is the general case for which duplicate is the special case consisting of two samples or measurements.

Routine method—A method used in recurring analytical problems.

Sample—A portion of a population or lot. It may consist of an individual or groups of individuals. It may refer to objects, materials, or measurements, conceivable as part of a larger group that could have been considered.

Secondary standard—A standard whose value is based upon comparison with some primary standard. Note that a secondary standard, once its value is established, can become a primary standard for some other user.

Segment—A specifically demarked portion of a lot, either actual or hypothetical.

Selectivity—The ability of methodology or instrumentation to respond to a desired substance or constituent and not to others. (See also Cross sensitivity.)

Sensitivity—Capability of methodology or instrumentation to discriminate between samples having differing concentrations or containing differing amounts of an analate.

Significant figure – A figure(s) that remains to a number or decimal after the ciphers to the right or left are cancelled.

Special cause – A cause of variance or bias that is external (not inherent) to the measurement system.

Split sample – A replicate portion or subsample of a total sample obtained in such a manner that it is not believed to differ significantly from other portions of the same sample.

Standard – A substance or material with properties believed to be known with sufficient accuracy to permit its use to evaluate the same property of another. In chemical measurements, it often describes a solution or substance commonly prepared by the analyst to establish a calibration curve or the analytical response function of an instrument.

Standardization – The process whereby the value of a potential standard is fixed by measurement with respect to a standard(s) of known value.

Standard addition – A method in which small increments of a substance under measurement are added to a sample under test to establish a response function, or to determine by extrapolation the amount of a constituent originally present in the test sample.

Standard method – A method (or procedure) of test developed by a standards-writing organization, based on consensus opinion or other criteria and often evaluated for its reliability by a collaborative testing procedure.

Standard operations procedure (SOP) – A procedure adopted for repetitive use when performing a specific measurement or sampling operation. It may be a standard method or one developed by the user.

Standard reference material – A reference material distributed and certified by the National Bureau of Standards.

Strata – Segments of a lot that may vary with respect to the property under study.

Subsample – A portion taken from a sample. A laboratory sample may be a subsample of a gross sample; similarly, a test portion may be a subsample of a laboratory sample.

Technique – A physical or chemical principle utilized separately or in combination with other techniques to determine the composition (analysis) of materials.

Test portion (also called specimen, test specimen, test unit, aliquot)—That quantity of a material of proper size for measurement of the property of interest. Test portions may be taken from the gross sample directly, but often preliminary operations, such as mixing or further reduction in particle size, are necessary.

Tolerance interval—That range of values, calculated from an estimate of the mean and standard deviation, within which a specified percentage of individual values of a population (measurements or sample) are expected to lie with a stated level of confidence.

Traceability—The ability to trace the source of uncertainty of a measurement or a measured value.

Training—Formal or informal instruction designed to provide competence of a specific nature.

Uncertainty—The range of values within which the true value is estimated to lie. It is a best estimate of possible inaccuracy due to both random and systematic error.

Validation—The process by which a sample, measurement method, or a piece of data is deemed useful for a specified purpose.

Variance—The value approached by the average of the sum of the squares of deviations of individual measurements from the limiting mean. Mathematically, it may be expressed as:

$$\frac{\Sigma (X_i - m)^2}{n} \rightarrow \sigma^2 \text{ as } n \rightarrow \infty$$

Ordinarily, it cannot be known but only its estimate, s^2, which is calculated by the expression:

$$s^2 = \frac{\Sigma (X_i - \bar{X})^2}{n - 1}$$

Warning limits—The limits shown on a control chart within which most of the test results are expected to lie (within a 95% probability) while the system remains in a state of statistical control.

Youden plot—A graphical presentation of data, recommended first by W. J. Youden, in which the result(s) obtained by a laboratory on one sample is plotted with respect to the result(s) it obtained on a similar sample. It helps in deciding whether discrepant results are due to random or systematic error.

APPENDIX B

Quality Assurance Program Documentation

This appendix contains supplemental material which may be useful in the development of documents related to a quality assurance program. The material is generic and believed to be widely useful. However, each laboratory should consider its applicability and develop the specific content most useful to its own measurement program.

The material consists largely of outlines that can be followed when preparing specific documentation. It is recommended that a laboratory or organization that uses it should address each item in an outline and prepare text that is appropriate for its own situation.

An exception to the outline format is Appendix B.1 which contains a complete text for a program document. This text is presented as an example of such a document and is intended only for guidance. Undoubtedly, it will need to be modified to meet the needs of a specific laboratory.

APPENDIX B.1 EXAMPLE QUALITY ASSURANCE PROGRAM DOCUMENT

1. Policy

The objective of the measurement program of (name of organization) is to provide high-quality research and analytical measurement data which are accurate, reliable, and adequate for the intended purpose in a cost-effective manner. To this end, the management is dedicated to the encouragement of excellence in measurement and to provide the physical and mental environment conducive to its achievement. To further these objectives, the quality assurance program described in the following sections has been established and applies to all scientific and technical work conducted, in principle and in detail, to the extent possible and feasible.

2. Purpose

The purpose of this document is to formalize the quality assurance practices that have been developed and are established to guide the staff in the production of quality outputs. A further purpose is to inform users of the services of the conditions and practices related to the services and data provided.

3. General Aspects

3.1 Preliminary requirements

All analytical work will be done according to a plan individually selected and/or developed and optimized for each situation. The requirements of each problem will be studied and thoroughly understood beforehand, in order to assure that the measurements undertaken will be adequate in kind, number, and quality. All measurements are made using standard methods, methods having peer recognition, methods developed in this laboratory, or those mandated by legal requirements. No method will be used to obtain data until it is known to be applicable and competence has been acquired in its use. If or when it appears that available techniques are not adequate to solve a particular problem, the user will be so informed and advised of any research or investigation required to develop adequate methodology and the estimated cost. Demonstration and documentation of the attainment of statistical control is a prerequisite for reporting any data, whether of a routine or research nature.

All laboratory equipment shall be maintained in a reasonable and proper state of repair. Space shall be maintained according to rules of cleanliness, order, and efficiency to facilitate the measurements and to protect the health and safety of the staff. No equipment shall be used until it is ascertained that it is in a safe and reliable operational state, and then only by personnel who have been thoroughly trained and duly qualified as operators. Work done for any organization will adhere to any project plan or quality assurance plan that it may specify to the extent possible. Whenever such directives differ from ongoing policy, such differences will be reconciled.

3.2 Experimental Plan

All experimental work will be done according to a written plan developed in advance of the actual work. Where applicable and appropriate, the plan will be developed in consultation with the client and with the assistance of a statistician. The plan will include the sampling strategy, the sampling procedure, the calibration process, the measurement procedure, and the data handling and validation process. The quality assurance to be followed shall conform to the general practices of the laboratory and to specific practices pertinent to the measurements carried out. All procedures used shall be documented and available for review by clients. Standard Operations Procedures (SOPs) are used to the extent possible. Any procedure used is validated by the analyst before use.

3.3 Reagents

Reagents are defined as any substances used to dissolve, disperse, extract, react with, or dilute any sample or analytical component of the sample. Strict control of reagents shall be maintained to minimize contamination or degradation. Dilutions or solutions prepared from them shall be clearly labeled and dated. The shelf life shall be designated and strictly observed. The highest quality reagents shall be used in all analytical work. Records shall be kept of all reagents used in every critical operation and appropriate blank corrections shall be evaluated and made as necessary.

3.4 Laboratory Safety

General safety rules shall be enforced. Special safety rules as required by specific situations shall be established and followed.

3.5 Disposal of Toxic Substances

The potential hazard of each substance used regularly or occasionally shall be known. No hazardous substances will be used until such use is approved by management. Safe disposal practices will be followed.

4. Samples

It is recognized that sampling is one of the most critical steps in chemical measurement. All aspects of sampling, sample preparation, and subsampling will be done according to an approved plan. Sample integrity will be protected by an appropriate chain of custody. Any limitations on the analytical results due to the samples measured will be specified.

5. Analytical Methodology

All measurements shall be made using methodology appropriate for the specific purpose. Each method will be described in writing before use. All measurements in a sequence will utilize the same method. Any significant changes in a procedure will be documented, including the reasons for the changes. It shall be made clear whether the changes are permanent or only for a particular use. The written procedure will contain the following information as a minimum:

a. Sample preparation and treatment
b. Chemical operations
c. Calibration procedure
d. Measurement procedure
    1. Instrumental conditions
    2. Instrumental adjustments
    3. Critical tolerances that must be observed
e. Data validation procedure

5.1 Standard Operations Procedures

Preference will be given to using standard methods when applicable. All methods used will be converted to SOPs as feasible, in the format used by ASTM.

5.2 Calibration

5.2.1 Any equipment or methodology that is used to provide numerical data that can influence a measured value will be calibrated to the accuracy requirements for its use. Records shall be kept of all calibrations, either made externally or within the laboratory. Calibration schedules shall be established for all aspects of physical and chemical measurements and shall be strictly observed.

5.2.2 Physical standards and measuring devices shall have currently valid calibrations, traceable to national standards. Calibration records and certificates shall be filled in a central location, and corrections shall be applied as required by the accuracy of the data.

5.2.3 Chemical calibrations and standardizations shall be made using standards prepared by methods reflecting state of the analytical art and materials of known purity.

5.2.4 Calibrations and standards obtained externally must adhere to the requirements for internal standards. Suppliers must certify compliance and provide evidence of the quality of the services and materials provided on request.

5.3 Data Handling

Descriptions of every analytical method used shall include procedures for data handling. The following subjects shall be addressed as appropriate:

- calculations, corrections, adjustment to standard conditions, normalization of data, computer programs
- statistical procedures used to report data, plotting procedures, curve fitting procedures
- corrections for systematic errors
- checks for internal consistency

When any of the above are included in GLP documents, citations to these will satisfy the requirements.

6. Laboratory Records

6.1 General

It is the responsibility of each analyst to keep complete records of all work or operations performed in the format prescribed by the particular organizational unit. The minimum requirement shall be that records shall be understandable to one versed in the art yet not directly connected with the work described. All laboratory records are the property of the laboratory and may not be removed from the premises without permission of higher authority. Records are to be safeguarded and stored in conformance with policy. Records of work done for clients shall be safeguarded according to their requirements if these require special conditions not covered by laboratory policy. Unauthorized changes, loss, or destruction of records can be grounds for dismissal and/or criminal actions.

6.2 Measurement Data

All experimental data will be recorded in bound laboratory notebooks using a permanent ink. Transcriptions shall be avoided wherever possible. The data shall be suitably annotated with reference to experimental details, conclusions, and directions for further work as appropriate. Any corrections to recorded data shall be made by crossing out the original and substituting or referring to the new data. Reasons for changes must be included. Cross references to relevant data shall be made as pertinent. Data in the form of charts, instrument recordings, and printouts will be given suitable identification and reference made to them in notebooks.

In view of the possible legal use of much of the data produced, all records shall be maintained in such a way as to preclude its discreditation at any time. Records of possible use in patent situations shall be attested by witnesses.

6.3 Maintenance Records

Records of maintenance of equipment and facilities shall be suitably recorded. Routine maintenance may be indicated by suitable labels. Maintenance that results in significant modification of equipment must be described in detail and recorded in operational manuals for the specific item.

6.4 Equipment Manuals

Manuals shall be suitably filed for ready access as needed.

7. Control Charts

Control charts are recognized as a prime means to document the statistical control of the measurement process and to describe measurement proficiency. It is a matter of policy that each measurement competence shall develop and maintain suitable control charts for all critical operations and suboperations to the extent feasible. When the use of control charts and specific control samples are specified in a given protocol, they shall be maintained in as close to real time as possible and shall be the basis for corrective actions when indicated.

8. Quality Assessment

8.1 All data reported must include a statement of its uncertainty, and the work plan must include a means for the determination or assignment of such limits. SRMs, other reference materials, spikes, or confirmation by another measurement technique are ways that may be used for this purpose. Statistically established confidence limits and an analysis of sources of systematic error can be used in the absence of experimental demonstration of limits of inaccuracy. No matter what approach is used, it must be identified, and the calculations upon which the limits are based shall be a matter of record.

8.2 Any significant problems in the measurement process must be resolved before reporting, or the data must be qualified by a statement describing the problem and its implications.

9. Data Reviewing and Reporting

9.1 All data will be subject to review by the management chain before release. All releases must be in writing. Oral preliminary releases are prohibited unless prior permission is obtained from management and provided they are subsequently confirmed in writing (within 24 hours). The lowest level of review is by the supervisor or designee. Sensitive data requires higher level review and release. Publication of data is predicated on editorial review. Data obtained for a client or sponsor is considered to be his property, and its dissemination is subject to his release. Data of a highly routine nature will follow the spirit of these regulations with exceptions as appropriate.

9.2 Reports will be prepared in a format suited to the particular end use. Each report, as possible, will contain the following information:

a. Summary of the problem investigated
b. Complete description of samples
c. Description of methodology (or reference)
d. Summary of data and reference to original data
e. Interpretations and conclusions
f. Attestation

9.3 Signatures

The analyst(s) involved will sign reports as well as all who review them. All signors attest that the data and associated information contained in the report are believed to be correct and that all quality assurance requirements have been fulfilled unless exceptions are approved and noted.

10. Research Investigations

All research work will be carried out efficiently, effectively, and with a high degree of technical excellence. The principles of good research parallel those for good measurement and differ only in detail. Accordingly, the spirit of the rules and regulations set forth above apply to all investigative work done in the laboratory.

Research reports are subject to internal controls, with responsibilities delegated as follows:

Author/investigator: Accuracy of data, adequate support for all conclusions, full disclosure and description of all significant details, maintenance of appropriate records, safeguarding of all data.

Middle manager: Technical review of all work.

Upper management: Editorial and technical review. Release of reports. Reviews by third parties as deemed appropriate.

11. Audits

A system of semiannual internal audits is established to review the ongoing quality assurance practices for compliance with the quality assurance program. This shall be conducted by a committee appointed by the laboratory director and chaired by the quality assurance director. This committee shall report on the health of the program and make recommendations for corrective actions as required.

12. Implementation

Each and every member of the staff has a stake in the quality of the laboratory's work and has responsibility for the implementation of quality assurance, according to her/his technical and/or managerial responsibilities. Specific responsibilities have been mentioned elsewhere in this document. The general aspects are summarized below.

Individuals: It is the responsibility of each person to support the quality assurance program in principle and detail. Each staff member is responsible for the technical quality of her/his work. All work must be done with the highest level of integrity and professional competence. Each member must be ever alert to problems or sources of error that could compromise the quality of technical work.

Management Chain: Management has the responsibility for supervising and administering the quality assurance program and providing an environment in which quality work can be produced. It is their responsibility that all quality assurance requirements of clients or sponsors are strictly followed.

Quality assurance director: The quality assurance director provides an ongoing supervision of the program and reports to management on any actual or perceived problems. She/he periodically reviews the program and makes recommendations for its improvement. She/he serves as the focal point for quality assurance in dealings with clients and governmental agencies.

Approved: ______________________________
Date: ______________________________

APPENDIX B.2 OUTLINE FOR USE IN PREPARING AN SOP FOR A MEASUREMENT METHOD

The following outline is recommended for use when preparing an SOP for a method of measurement. It follows essentially the format used by ASTM, which is the basis for literally thousands of test methods that many laboratories use in their measurement programs. Adoption of the same format for their internal methods provides a consistency of style which facilitates their use.

SOP No. ______________________________

1. Title
2. Scope
   Parameters measured, range, matrix, expected precision and accuracy
3. Referenced documents
4. Terminology
5. Summary of method
6. Significance and use
7. Interferences
8. Apparatus
9. Reagents and materials
10. Hazards and precautions
11. Sampling, sample preparation
12. Preparation of apparatus
13. Calibration and standardization
14. Procedure
15. Demonstration of statistical control
16. Calculations
17. Assignment of uncertainty

APPENDIX B.3 OUTLINE FOR USE IN PREPARING AN SOP FOR CALIBRATION

The following outline may be found useful when preparing an SOP for the calibration of laboratory equipment or standards.

SOP No. ______________________

1. Title
2. Summary
3. Description of item calibrated
4. Calibration interval
5. Standards needed
   a. Source
   b. Preparation
6. Procedure
7. Calculations
   a. Reduction of data
   b. Curve fitting
   c. Uncertainty limits
8. Report
   a. Format
   b. Labeling and approval
9. References

APPENDIX B.4 OUTLINE FOR USE IN PREPARING A GLP OR GMP

The following outline is recommended for use when preparing a GLP or a GMP. The outline is very general, and the content of the practice will vary greatly with the subject.

GLP (or GMP) No. ______________________

1. Title
2. Introduction
3. Scope
4. Significance and use
5. Procedure(s)
6. References

APPENDIX C

Statistical Tables

This appendix contains a selection of statistical tables that are used most often when statistically analyzing measurement data. The tables are abbreviated forms of larger tables found in statistical books such as the National Bureau of Standards Handbook 91 [100]. While the tables included should be adequate for data sets ordinarily encountered in most laboratories, large data sets will require the more extensive listings cited in each table.

Table C.1. Use of Range to Estimate Standard Deviation

Number of Sets of Replicates	Factor Degrees of Freedom	Number of Replicates in Set				
		2	3	4	5	6
	d^*_2	1.41	1.91	2.24	2.48	2.67
1	df	1.00	1.98	2.93	3.83	4.68
	d^*_2	1.23	1.77	2.12	2.38	2.58
3	df	2.83	5.86	8.44	11.1	13.6
	d^*_2	1.19	1.74	2.10	2.36	2.56
5	df	4.59	9.31	13.9	18.4	22.6
	d^*_2	1.16	1.72	2.08	2.34	2.55
10	df	8.99	18.4	27.6	36.5	44.9
	d^*_2	1.15	1.71	2.07	2.34	2.54
15	df	13.4	27.5	41.3	54.6	67.2
	d^*_2	1.14	1.70	2.07	2.33	2.54
20	df	17.8	36.5	55.0	72.7	89.6
∞	d^*_2	1.13	1.69	2.06	2.33	2.53

Intermediate values for d^*_2 and df may be obtained by interpolation, or from the reference from which this table was adapted. Adapted from Lloyd S. Nelson. *J. Qual. Tech. 7* No. 1. January 1975.

Table C.2. Z-Factors for Two-Sided Confidence Intervals for the Normal Distribution

Confidence Level, %	Z-Factor
50	0.68
67	1.00
75	1.15
90	1.645
95	1.960
95.28	2.000
99	2.575
99.74	3.000
99.9934	4.000
99.99995	5.000
$100 - 10^{-9}$	6.000

Table C.3 Student t Variate

*	80%	90%	95%	98%	99%	99.73% Z=3
df	$t_{.90}$	$t_{.95}$	$t_{.975}$	$t_{.99}$	$t_{.995}$	$t_{.9985}$
1	3.078	6.314	12.706	31.821	63.657	235.80
2	1.886	2.920	4.303	6.965	9.925	19.207
3	1.638	2.353	3.182	4.541	5.841	9.219
4	1.533	2.132	2.776	3.747	4.604	6.620
5	1.476	2.015	2.571	3.365	4.032	5.507
6	1.440	1.943	2.447	3.143	3.707	4.904
7	1.415	1.895	2.365	2.998	3.499	4.530
8	1.397	1.860	2.306	2.896	3.355	4.277
9	1.383	1.833	2.262	2.821	3.250	4.094
10	1.372	1.812	2.228	2.764	3.169	3.975
11	1.363	1.796	2.201	2.718	3.106	3.850
12	1.356	1.782	2.179	2.681	3.055	3.764
13	1.350	1.771	2.160	2.650	3.012	3.694
14	1.345	1.761	2.145	2.624	2.977	3.636
15	1.341	1.753	2.131	2.602	2.947	3.586
16	1.337	1.746	2.120	2.583	2.921	3.544
17	1.333	1.740	2.110	2.567	2.898	3.507
18	1.330	1.734	2.101	2.552	2.878	3.475
19	1.328	1.729	2.093	2.539	2.861	3.447
20	1.325	1.725	2.086	2.528	2.845	3.422
25	1.316	1.708	2.060	2.485	2.787	3.330
30	1.310	1.697	2.042	2.457	2.750	3.270
40	1.303	1.684	2.021	2.423	2.704	3.199
60	1.296	1.671	2.000	2.390	2.660	3.130
∞	1.282	1.645	1.960	2.326	2.576	3.000

*Columns to be used in calculating corresponding two-sided confidence interval. Excerpted from "Experimental Statistics," NBS Handbook 91 [100]. Last column from B. J. Joiner, *J. Research* NBS.

Table C.4. Factors for Computing Two-Sided Confidence Intervals for σ

Degrees of Freedom	$\alpha = .05$		$\alpha = .01$		$\alpha = .001$	
	B_U	B_L	B_U	B_L	B_U	B_L
1	17.79	.358	86.31	.297	844.4	.248
2	4.86	.458	10.70	.388	33.3	.329
3	3.18	.518	5.45	.445	11.6	.382
4	2.57	.559	3.89	.486	6.94	.422
5	2.25	.590	3.18	.518	5.08	.453
6	2.05	.614	2.76	.544	4.13	.478
7	1.92	.634	2.50	.565	3.55	.500
8	1.82	.651	2.31	.583	3.17	.519
9	1.75	.666	2.17	.599	2.89	.535
10	1.69	.678	2.06	.612	2.69	.549
15	1.51	.724	1.76	.663	2.14	.603
20	1.42	.754	1.61	.697	1.89	.640
25	1.36	.775	1.52	.721	1.74	.667
30	1.32	.791	1.46	.740	1.64	.688
35	1.29	.804	1.41	.755	1.58	.705
40	1.27	.815	1.38	.768	1.52	.720
45	1.25	.824	1.35	.779	1.48	.732
50	1.23	.831	1.33	.788	1.45	.743

Excerpted from "Experimental Statistics," NBS Handbook 91 [100].

Table C.5. Factors for Computing Two-Sided Tolerance Intervals for a Normal Distribution

	$\tau = 0.95$				$\tau = 0.99$			
n/p	**.90**	**.95**	**.99**	**.999**	**.90**	**.95**	**.99**	**.999**
2	32.02	37.67	48.43	60.57	160.19	188.49	242.30	303.05
3	8.38	9.92	12.86	16.21	18.93	22.40	29.06	36.62
4	5.37	6.37	8.30	10.50	9.40	11.15	14.53	18.38
5	4.28	5.08	6.63	8.42	6.61	7.86	10.26	13.02
6	3.71	4.41	5.78	7.34	5.34	6.34	8.30	10.55
7	3.37	4.01	5.25	6.68	4.61	5.49	7.19	9.14
8	3.14	3.73	4.89	6.23	4.15	4.94	6.47	8.23
9	2.97	3.53	4.63	5.90	3.82	4.55	5.97	7.60
10	2.84	3.38	4.43	5.65	3.58	4.26	5.59	7.13
15	2.48	2.95	3.88	4.95	2.94	3.51	4.60	5.88
20	2.31	2.75	3.62	4.61	2.66	3.17	4.16	5.31
25	2.21	2.63	3.46	4.41	2.49	2.97	3.90	4.98

Excerpted from "Experimental Statistics," NBS Handbook 91 [100].

Table C.6. Critical Values for the F Test, $F_{0.975}$

	df_N									
df_D	**1**	**2**	**4**	**6**	**8**	**10**	**15**	**20**	**30**	**40**
1	648	800	900	937	957	969	985	993	1001	1006
2	38.5	39.0	39.2	39.3	39.4	39.4	39.4	39.4	39.5	39.5
4	12.2	10.6	9.6	9.2	9.0	8.8	8.7	8.6	8.5	8.4
6	8.8	7.3	6.2	5.8	5.6	5.5	5.3	5.2	5.1	5.0
8	7.6	6.1	5.0	4.6	4.4	4.3	4.1	4.0	3.9	3.8
10	6.9	5.5	4.5	4.1	3.8	3.7	3.5	3.4	3.3	3.3
15	6.2	4.8	3.8	3.4	3.2	3.1	2.9	2.8	2.6	2.6
20	5.9	4.5	3.5	3.1	2.9	2.8	2.6	2.5	2.4	2.3
30	5.6	4.2	3.2	2.9	2.6	2.5	2.3	2.2	2.1	2.0
40	5.4	4.0	3.1	2.7	2.5	2.4	2.2	2.1	1.9	1.9
60	5.3	3.9	3.0	2.6	2.4	2.3	2.1	1.9	1.8	1.7
120	5.2	3.8	2.9	2.5	2.3	2.2	1.9	1.8	1.7	1.6
∞	5.0	3.7	2.8	2.4	2.2	2.1	1.8	1.7	1.6	1.5

For use for a one-tailed test of equality of standard deviation estimates at 2.5% level of confidence, or for a two-tailed test at 5% level of confidence. df_N and df_D refer to degrees of freedom of variances of numerator and denominator of F test, respectively. Excerpted from "Experimental Statistics," NBS Handbook 91 [100].

Table C.7. Values for Use in the Dixon Test for Outliers

Statistic	Number of Observations, n	Risk of False Rejection 0.5%	1%	5%	10%
	3	.994	.988	.941	.886
	4	.926	.889	.765	.679
r_{10}	5	.821	.780	.642	.557
	6	.740	.698	.560	.482
	7	.680	.637	.507	.434
	8	.725	.683	.554	.479
r_{11}	9	.677	.635	.512	.441
	10	.639	.597	.477	.409
	11	.713	.679	.576	.517
r_{21}	12	.675	.642	.546	.490
	13	.649	.615	.521	.467
	14	.674	.641	.546	.492
	15	.647	.616	.525	.472
	16	.624	.595	.507	.454
r_{22}	17	.605	.577	.490	.438
	18	.589	.561	.475	.424
	19	.575	.547	.462	.412
	20	.562	.535	.450	.401

Tabulated values obtained from Natrella [100]. See page 36 for discussion. Original reference: W. J. Dixon, "Processing Data Outliers," Biometrics, BIOMA, 9 (No.1): 74–89 (March 1953).

Table C.8. Values for Use in the Grubbs Test for Outliers

Number of	Risk of False Rejection				
Data Points	0.1%	0.5%	1%	5%	10%
3	1.155	1.155	1.155	1.153	1.148
4	1.496	1.496	1.492	1.463	1.425
5	1.780	1.764	1.749	1.672	1.602
6	2.011	1.973	1.944	1.822	1.729
7	2.201	2.139	2.097	1.938	1.828
8	2.358	2.274	2.221	2.032	1.909
9	2.492	2.387	2.323	2.110	1.977
10	2.606	2.482	2.410	2.176	2.036
15	2.997	2.806	2.705	2.409	2.247
20	3.230	3.001	2.884	2.557	2.385
25	3.389	3.135	3.009	2.663	2.486
50	3.789	3.483	3.336	2.956	2.768
100	4.084	3.754	3.600	3.207	3.017

Tabulated values obtained in part from ASTM E-178 [15] which should be consulted for more extensive tables. See page 37 for discussion of treatment of outliers. Original reference: F. E. Grubbs and G. Beck, "Extension of Sample Sizes and Percentage Points for Significance Tests of Outlying Observations," Technometrics, TCMTA, 14 (No. 4): 847–54 (November 1972).

Table C.9. Values for Use in the Youden Test to Identify Consistent Outlying Performance

Approximate 5% Two-Tail Limits for Ranking Scores*

Number of Partici-pants	Number of Materials												
	3	**4**	**5**	**6**	**7**	**8**	**9**	**10**	**11**	**12**	**13**	**14**	**15**
3		4	5	7	8	10	12	13	15	17	19	20	22
		12	15	17	20	22	24	27	29	31	33	36	38
4		4	6	8	10	12	14	16	18	20	22	24	26
		16	19	22	25	28	31	34	37	40	43	46	49
5		5	7	9	11	13	16	18	21	23	26	28	31
		19	23	27	31	35	38	42	45	49	52	56	59
6	3	5	7	10	12	15	18	21	23	26	29	32	35
	18	23	28	32	37	41	45	49	54	58	62	66	70
7	3	5	8	11	14	17	20	23	26	29	32	36	39
	21	27	32	37	42	47	52	57	62	67	72	76	81
8	3	6	9	12	15	18	22	25	29	32	36	39	43
	24	30	36	42	48	54	59	65	70	76	81	87	92
9	3	6	9	13	16	20	24	27	31	35	39	43	47
	27	34	41	47	54	60	66	73	79	85	91	97	103
10	4	7	10	14	17	21	26	30	34	38	43	47	51
	29	37	45	52	60	67	73	80	87	94	100	107	114
11	4	7	11	15	19	23	27	32	36	41	46	51	55
	32	41	49	57	65	73	81	88	96	103	110	117	125
12	4	7	11	15	20	24	29	34	39	44	49	54	59
	35	45	54	63	71	80	88	96	104	112	120	128	136
13	4	8	12	16	21	26	31	36	42	47	52	58	63
	38	48	58	68	77	86	95	104	112	121	130	138	147
14	4	8	12	17	22	27	33	38	44	50	56	61	67
	41	52	63	73	83	93	102	112	121	130	139	149	158
15	4	8	13	18	23	29	35	41	47	53	59	65	71
	44	56	67	78	89	99	109	119	129	139	149	159	169

* Approximate, because of rounding. From W.J. Youden [81].

Table C.10. Values for Use in the Cochran Test for Extreme Values for Variance

(Five percent risk of wrong decision)

Number of Variances Compared	Number of Replicate Values Used to Compute Each Variance							
	2	3	4	5	6	7	10	∞
2	.9985	.9750	.9392	.9057	.8772	.8534	.8010	.5000
3	.9969	.8709	.7977	.7457	.7071	.6771	.6167	.3333
4	.9065	.7679	.6841	.6287	.5895	.5598	.5017	.2500
5	.8412	.6838	.5981	.5441	.5065	.4783	.4214	.2000
6	.7808	.6161	.5321	.4803	.4447	.4184	.3682	.1667
7	.7271	.5612	.4800	.4307	.3974	.3726	.3259	.1429
10	.6020	.4450	.3733	.3311	.3029	.2823	.2439	.1000
20	.3894	.2705	.2205	.1921	.1735	.1602	.1357	.0500
30	.2929	.1980	.1593	.1377	.1237	.1137	.0958	.0333
40	.2370	.1576	.1259	.1082	.0968	.0887	.0745	.0250
60	.1737	.1131	.0895	.0765	.0682	.0623	.0520	.0167

Extract from Table 14.1, Eisenhart, Hastay, and Wallis, Selected Techniques of Statistical Analysis, McGraw-Hill Book Co.(1947). Each variance must be estimated with the same number of degrees of freedom. Calculate s^2 (largest)/$\Sigma\ s^2_i$. If ratio exceeds tabulated value, assume largest value is extreme with 95% confidence (5% risk of wrong decision).

Table C.11. Table of Random Numbers

57	00	62	10	37	53	46	70	15	00	84	24	45	99	08	48
42	30	47	95	79	97	49	26	03	29	15	40	42	38	41	82
38	62	52	51	08	32	97	84	94	28	91	67	18	86	57	87
71	53	56	07	00	37	36	31	22	33	59	90	76	82	58	42
50	08	48	24	64	29	85	26	95	46	57	88	14	91	61	56
58	40	89	00	68	73	04	39	32	82	88	07	38	93	71	90
98	41	39	34	35	31	64	66	10	98	51	08	86	27	94	44
43	28	56	35	51	53	39	49	01	51	38	50	60	36	91	17
22	73	18	28	64	56	91	48	20	81	17	05	74	25	86	10
59	30	99	06	28	61	94	60	69	98	28	04	85	35	93	74
35	17	25	93	05	57	75	27	56	00	31	07	34	06	39	30
83	79	77	21	47	01	24	86	06	60	70	67	22	49	89	32
03	33	81	79	88	24	53	97	44	39	37	60	78	68	71	69
71	37	34	24	07	84	39	21	91	58	78	11	31	09	30	00
49	15	18	28	02	93	01	49	33	09	07	22	78	73	78	13
60	09	34	02	36	47	72	09	24	05	16	41	05	54	67	24
45	20	63	89	09	12	00	80	75	57	70	38	97	37	88	20
03	44	96	98	71	63	28	14	62	12	28	47	10	56	15	39
36	01	74	57	36	17	01	64	45	18	73	38	84	82	34	71
92	15	21	00	05	54	99	23	27	73	62	64	14	81	46	88
85	45	88	22	62	68	91	76	61	20	72	44	75	96	28	60
33	06	71	54	80	35	19	69	33	50	19	55	92	73	47	18
02	29	09	54	25	92	15	60	61	22	30	43	68	47	26	24
40	06	34	46	66	19	81	48	65	87	71	25	96	66	00	39
15	16	48	77	54	14	02	48	90	72	75	65	29	86	02	56
59	99	11	83	74	42	89	14	37	71	56	43	21	11	06	40
32	02	23	48	65	95	13	05	03	18	07	04	09	60	21	65
81	24	55	38	16	43	54	08	84	68	44	31	91	43	28	58
79	51	90	44	28	88	40	84	72	85	06	50	56	69	96	67
15	08	29	25	46	86	48	21	17	86	65	37	03	69	63	09

(cont.)

Table C.11. Continued

13	58	60	84	97	72	09	40	61	99	49	67	22	96	76	26
78	16	78	96	33	19	83	90	14	07	86	32	83	27	57	89
95	51	20	08	29	36	33	99	67	29	16	24	61	52	47	97
06	44	95	29	75	68	17	24	47	25	67	20	38	90	87	40
03	47	42	56	97	43	87	57	72	36	68	38	83	28	09	53
97	01	74	72	80	89	39	24	19	16	22	24	67	69	26	09
40	13	80	72	66	57	07	67	25	93	27	74	70	62	77	17
71	96	32	12	14	53	41	11	26	03	98	90	04	72	73	45
28	32	24	78	23	18	73	26	21	86	85	12	61	14	67	60
16	34	83	15	56	49	11	53	66	05	18	96	30	69	09	77
43	65	78	71	04	59	35	20	77	50	97	38	25	28	98	10
31	81	42	15	40	93	15	53	08	79	02	50	15	24	19	27

APPENDIX D

Study Aids

This appendix consists of three sections containing material included to assist the reader who is using the book for comprehensive study of the discipline of quality assurance. Appendix D.1 and Appendix D.2 cover the terminology and most of the basic concepts of quality assurance. The reader may wish to test her/his knowledge by a periodic self-examination. The chapter headings and index may then be consulted to see if the "right answers" were given to the respective questions. Appendix D.3 consistes of 21 problems, typical of the various situations encountered when evaluating measurement data. The answers are not given, but similar examples are discussed in the text. Appendix D.4 consists of a template that the author has found to be useful when describing various aspects of quality assurance and when analyzing measurement problems.

APPENDIX D.1 SEVENTY-FIVE TERMS COMMONLY USED IN DISCUSSING QUALITY ASSURANCE

Define or Describe:

1. quality, design quality, conformance quality
2. quality assurance
3. quality control
4. quality assessment
5. data set
6. sample statistics
7. population statistics
8. variance
9. degrees of freedom
10. z-factor
11. student's t
12. coefficient of variation
13. relative standard deviation
14. distribution
15. normal distribution
16. skewness
17. kurtosis
18. median
19. geometric mean
20. randomness
21. random number table
22. t-test
23. F-test
24. random sample
25. representative sampling
26. discriminatory sample
27. stratification
28. bulk sampling
29. split sample
30. foreign objects (sampling)
31. chain of custody
32. technique
33. method
34. procedure
35. protocol
36. measurement
37. definitive method
38. absolute method
39. figures of merit
40. sensitivity
41. selectivity
42. ruggedness
43. expertise
44. competence
45. reference method
46. calibration
47. standardization
48. primary and secondary standards
49. calibrant
50. intercalibration
51. joint confidence interval ellipse
52. linear relationship
53. spike; surrogate
54. recovery
55. blind, double blind
56. aliquot
57. Youden plot
58. outlier
59. range
60. residual plot
61. empirical
62. theoretical
63. blunder
64. technically sound
65. legally defensible
66. standard addition
67. limit of detection
68. limit of quantitation
69. limit of linearity
70. Pareto analysis
71. length-of-run concept
72. assignable cause, chance cause
73. units and standards
74. probability
75. quality assurance manual

APPENDIX D.2 SEVENTY-FIVE TOPICS RELATED TO QUALITY ASSURANCE OF MEASUREMENTS

1. Why is it necessary to know data quality?
2. How can one describe data quality?
3. Discuss how measurement uncertainty affects decisions.
4. Quality assessment.
5. Discuss chemical analysis as a system.
6. Discuss accuracy, precision, and bias.
7. Discuss statistical control and its application in the measurement system.
8. Discuss the basic prerequisites for applying statistics to data.
9. Discuss distributions. What is meant by a normal distribution and how could it be verified?
10. Describe three ways to estimate a standard deviation. What is meant by a pooled standard deviation and how is it calculated?
11. Discuss the minimum amount of statistical information needed to describe a data set.
12. Discuss the concept of a confidence interval for a mean. How is it calculated and what does it mean?
13. Discuss the concept of a confidence interval for sigma. How is it calculated and what does it mean?
14. Discuss the concept of a statistical tolerance interval for a population. How is it calculated and what does it mean?
15. How could one decide whether a measured value differs significantly from a standard (expected) value?
16. How could one decide whether an estimate of sigma differs significantly from an expected value?
17. How could one decide whether two estimates of sigma differ significantly?
18. How could one decide whether two measured values differ significantly?
19. Discuss the concept of outliers and ways to identify and confirm their existence.
20. Define the logical steps to combine data sets.
21. What is a model of a measurement problem and what is its importance?
22. Discuss the planning of a measurement program and its significance.
23. Discuss intuitive sampling; protocol sampling; statistical sampling.
24. Discuss sampling from these viewpoints: importance; kinds of samples; subsampling; sample preparation; holding time; sources of uncertainty; chain of custody; quality assurance.
25. Discuss sample homogeneity and its significance. How could you estimate sample variance? How could you use this information?
26. Discuss the guidance available for decisions on requirements for number of samples, size of samples, and number of measurements of samples. What are the options when requirements for the above are not feasible?

27. Discuss the need and logic for selecting appropriate methodology for a specific application.
28. How could one evaluate the precision and bias of a method of analysis?
29. Discuss the function and significance of collaborative testing.
30. Discuss detection, quantitation, and their limits.
31. Discuss calibration from the following viewpoints: importance; basic requirements; frequency; accuracy of standards; uncertainties; curve fitting.
32. What is quality control and what are its basic elements?
33. Discuss good laboratory practices (GLPs).
34. Discuss good measurement practices (GMPs).
35. Discuss standard operations procedures (SOPs).
36. Discuss protocols for specific purposes (PSPs).
37. Discuss the relation of education and training to QA.
38. Discuss the role of inspection in QA.
39. Discuss the importance of documentation in QA.
40. Discuss the basic philosophy of control charts from the viewpoints of: what; why; kinds; mechanics; use.
41. Discuss R-control charts from the viewpoints of: what; importance; kinds; advantages and limitations.
42. Discuss control limits: how to establish; when to revise; interpretation of control chart data.
43. Discuss quality assessment from the viewpoints of: what it is; various approaches; internal; external; frequency; corrective actions.
44. Discuss blanks from the viewpoints of: importance; kinds; sources; evaluation; control; correction; statistical considerations.
45. How can one identify sources of variance and quantify them? What is meant by long- and short-term standard deviations? What is the significance of each?
46. Discuss systems audits for QA.
47. Discuss performance audits for QA.
48. Discuss the concept and use of Ishikawa's cause-effect diagram.
49. Discuss the relation of personnel performance to quality data.
50. How could one decide whether one had excessive random error and what could be done about it?
51. How could one decide whether one had excessive measurement bias and what could be done about it?
52. Discuss the propagation of random errors.
53. Discuss the propagation of systematic (bias) errors.
54. Discuss and contrast: confidence limits; bounds for bias; uncertainty; errors.
55. Discuss quality circles from these viewpoints: what they are; how they operate; benefits and disadvantages; what they can accomplish. Suggest a list of topics that quality circles could address in your organization. Do you think they would be worthwhile? Justify your opinion.

56. What is meant by measurement compatibility? What is the function of national standards? What is meant by a national measurement system?
57. What is meant by traceability?
58. What are reference materials and how can they be utilized in a measurement process?
59. What are minimum prerequisites for reporting data?
60. Discuss various factors that could limit the usefulness of data.
61. Discuss the assignment of confidence limits and limits of uncertainty to data.
62. Discuss significant figures and the rounding off of data.
63. What is meant by validation of samples, methodology, and data?
64. Describe the various approaches and their relative merits for validation of a methodology for a specific application.
65. What is meant by laboratory certification and accreditation and what are the basic approaches used?
66. How could you evaluate your own laboratory or another laboratory for its general and specific capability?
67. Discuss the various general approaches to quality assurance and their relative merits and applicability.
68. Discuss the cost and benefit relations of a quality assurance program.
69. Discuss and contrast quality assurance of: large/limited-scope measurement programs; routine/research situations; individual/organizational QA.
70. Discuss logical approaches to the development of credible QA programs.
71. Discuss mandatory quality assurance.
72. Discuss quality assurance as a philosophy and as a system.
73. How can one "sell" quality assurance to management; staff; clients?
74. Discuss the management of quality assurance.
75. What is meant by a quality assurance code of ethics? What are the roles of management and staff?

APPENDIX D.3 QUALITY ASSURANCE PROBLEMS

The following problems are typical of many situations encountered when analyzing chemical measurement data. Units of measurement are omitted in most cases for convenience. However, it should be remembered that standard deviations and their estimates have the same units as the measurements that they represent and should be stated when reporting data. Only the coefficient of variation and the relative standard deviation are dimensionless.

In the problems, x_f and x_s mean the first and second measurements in a set, respectively.

Problem 1. Calculate estimates, s, of the standard deviation and the associated number of degrees of freedom for the following sets of data by three ways, as possible.

(1.1)

15.1
13.3

(1.2)

15.2
14.7
15.0
15.3
15.2
14.9
15.1

(1.3) Duplicate measurements

Set no.	x_f	x_s
1	14.7	15.0
2	15.1	14.9
3	15.0	15.1
4	14.9	14.9
5	15.3	14.8
6	14.9	15.1
7	14.9	15.0

(1.4) Duplicate measurements

Set no.	x_f	x_s
1	14.7	15.2
2	20.1	19.8
3	12.5	13.0
4	23.6	23.3
5	15.1	14.9
6	18.2	18.0
7	16.1	16.1

(1.5) Calculate the coefficient of variation and the relative standard deviation for each estimate of σ.

Problem 2. Estimates of the standard deviation were made on several occasions with the results shown. Calculate the pooled standard deviation and indicate the number of degrees of freedom associated with the result.

s	n
1.5	3
2.1	2
1.5	9
2.0	10
1.8	3

Problem 3. Calculate the means and the 95% confidence intervals for the following data sets:

(3.1)		(3.2)	
	15.0		15.0
	14.7		15.1
	14.9		
	15.0		
	14.8		
	14.7		
	15.0		

Problem 4. Laboratories A and B reported the following results for measurements of the same material. Is there any significant difference between the means and the standard deviation estimates of the two sets of measurements?

Laboratory A	Laboratory B
14.7	14.4
15.1	15.0
14.9	14.9
15.0	14.8
14.8	15.1
15.3	14.7

Problem 5. A client's sample of fuel oil was analyzed for its S content with the following results:

2.01% S, 2.03% S

The maximum allowable sulfur content is 2.00% S. Does the fuel oil meet the specification? How would you discuss this with the client?

Problem 6. SRM 1621, Sulfur in Fuel Oil, was analyzed with the following results:

0.95% S , 0.97% S . 0.96% S

The certified value is 0.950 ± .005% S. What is your conclusion about the possible bias of the method used?

Problem 7. The following sets of data were taken to evaluate a measurement procedure and to establish a control chart.

No.	Set 1		Set 2		Set 3	
	X_f	X_s	X_f	X_s	X_f	X_s
1	2.0	1.9	20.0	21.0	50.7	47.0
2	1.7	2.3	21.3	19.9	54.3	51.3
3	2.4	2.0	22.5	20.3	49.8	50.9
4	2.4	2.4	19.9	20.0	53.3	51.2
5	1.9	2.1	21.0	22.0	53.0	52.0
6	2.1	2.0	20.5	19.7	51.0	53.0
7	1.6	1.9	20.3	21.3	50.2	52.0

(7.1) Construct R control charts and an R performance chart.

(7.2) Calculate the sensitivity of the method when analyzing samples with concentrations of 10 and 40 units of analyte, respectively.

(7.3) Estimate the MDL.

(7.4) A sample was measured in duplicate with the following results: 13.7, 15.0. Use the control chart to help decide whether the results can be reported. If so, what result would you report? What assumptions are involved?

(7.5) What uncertainty would you assign to a single measurement made at the same time as (7.4) that gave the result 29.2?

(7.6) When using the above control charts, you made a single measurement of a sample and obtained the result 14.1. Another laboratory reported a value of 16.0 for the same sample. On questioning, you learned that its result was the mean of the following set of measurements: 15.7, 17.0, 15.3. Do the two results differ significantly?

Problem 8. Decide whether there are any outliers in the following sets of data:

Set 1	Set 2	Set 3
15.0	17.2	27.5
14.3	19.6	27.3
15.0	18.3	27.4
	14.2	27.7
	18.6	27.6
	19.0	27.1
		28.2
		26.9
		27.3
		27.2

Use more than one test, as appropriate, to support your decision.

Problem 9. A group of laboratories were each given samples of the same four materials to analyze. The following results were reported. Are any of the individual results outliers? Are any of these laboratories outliers?

	Results, mg/Liter on sample numbers			
Laboratory No.	1	2	3	4
A	.70	2.33	5.43	8.60
B	.75	2.40	5.60	8.70
C	.63	2.47	5.47	8.73
D	.55	2.05	4.95	8.29
E	.57	2.53	6.03	9.77
F	.65	2.70	6.57	10.50
G	.52	2.03	5.00	8.07

Problem 10. Seven samples were randomly selected from a batch of material and analyzed by duplicate measurements. Estimate the sample mean and the standard deviation of the samples.

Sample no.	X_f	X_s
1	15.0	15.2
2	13.9	14.3
3	14.8	14.8
4	15.1	14.9
5	14.7	14.2
6	14.9	14.7
7	14.1	14.4

Problem 11. Calculate the tolerance interval for coverage of 95% of the samples of Problem 10 with a confidence of 95%.

Problem 12. The following represents results of measurements on 15 occasions of an SRM with a certified value of 3.50.

3.50 3.48 3.51 3.52 3.49 3.50 3.55 3.49 3.51
3.52 3.54 3.49 3.55 3.50 3.53

Is there reason to believe the methodology is biased?

Problem 13. Establish an X control chart for use with the SRM of Problem 12, using the data of Problem 12.

Problem 14. When the control chart of Problem 13 was used, the next series of control sample measurements was as follows:

3.49, 3.50, 3.53, 3.50 3.49, 3.51, 3.50

Did any of these values indicate that the system was out of control? Has the precision changed significantly? Modify the control limits in an appropriate manner.

Problem 15. The next 20 measurements using the control chart of Problem 14 were as follows:

3.51, 3.52, 3.48, 3.46, 3.52, 3.50, 3.48, 3.48, 3.46, 3.44, 3.50,
3.48, 3.52, 3.50, 3.48, 3.50, 3.46, 3.52, 3.50, 3.52

What would you conclude from this set of data?

Problem 16. A client gave you two samples to analyze in order to characterize his production output. You used a procedure known to be in statistical control with a relative standard deviation of 3% and obtained the values 15.0 and 16.0, representing the averages of duplicate measurements on the respective samples. What would you report for the average concentration of the analyte in the production lot and the 95%, 95% tolerance interval for the lot?

Problem 17. The results reported in Problem 16 disturbed the client, so he submitted 10 new samples and asked for single measurements on each by the same procedure. The following results were then obtained:

16.1, 15.3, 15.8, 15.7, 15.2, 15.9, 15.4, 15.1, 15.6, 15.3

What could you say now about the variability of his production lot?

Problem 18. Calculate 95%, 99%, and 99.9% confidence intervals for each of following estimates of standard deviations:

s = 2.15 df = 2
s = 1.75 df = 5
s = 2.00 df = 10
s = 2.05 df = 20

Problem 19. Joe claimed that the method he uses is more precise than John's method. They decided on a contest to settle the question but agreed that a 95% level of significance would be used as a basis for decision. Each analyzed the same material five times with results for s as follows: Joe—1.50; John—2.11. What was the conclusion? What would have been the decision if the results were Joe—1.40; John—2.25? If an apparent difference in precision of a factor of 2 were true, how many measurements would need to be made to confirm it with 95% confidence?

Problem 20. On the assumption that a homogeneous material is analyzed, using a method with a relative standard deviation of 1%, how many measurements would be required if the confidence interval of the mean is desired to be within ± 0.05% ?

Problem 21. The relative standard deviation of a group of environmental samples is estimated to be ± 50%. The analytical method proposed for use has a relative standard deviation of ± 30%. How many samples and how many measurements of each would you propose if the relative error of the mean value is desired to be within ± 25% ?

APPENDIX D.4 PROBLEM IDENTIFICATION TEMPLATE

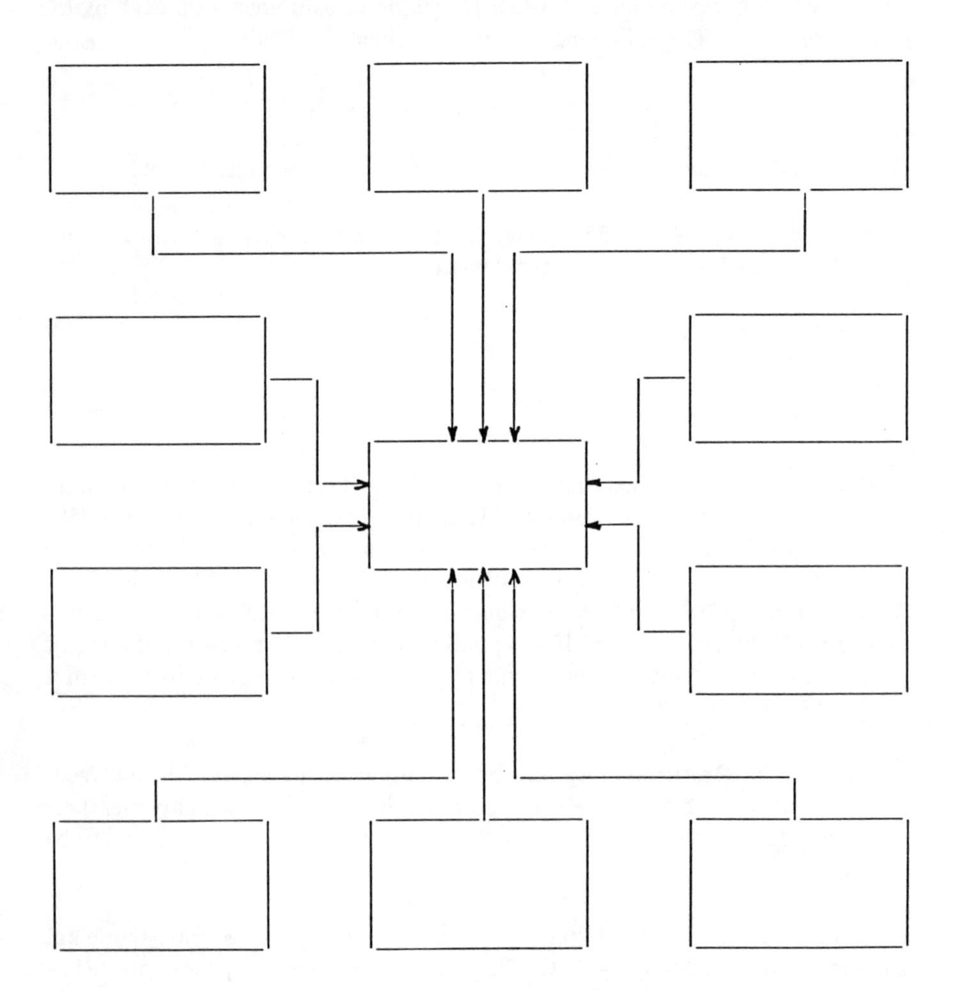

APPENDIX E

Annotated Bibliography of Selected Publications on Quality Assurance of Measurements

[1] American Council of Independent Laboratories (ACIL), "Quality Control System—Requirements for a Testing and Inspection Laboratory," American Council of Independent Laboratories, 1725 K St. NW, Washington, DC 20006.

A booklet describing the basic elements of quality control for a testing laboratory.

[2] American Chemical Society Committee Report, "Guidelines for Data Acquisition and Data Quality Evaluation in Environmental Chemistry," *Anal. Chem.* 52: 2242 (1980).

A number of items are discussed that should be considered when analyzing environmental samples. Limits of uncertainty must be assigned. Full attention must be given to quality assurance.

[3] American Chemical Society Committee Report, "Principles of Environmental Measurement," *Anal. Chem.* 55: 2210–2218 (1983).

This is a revision of the "Guidelines for Data Acquisition and Data Quality Evaluation in Environmental Chemistry" [2]. The original subject matter is considerably enlarged with emphasis on what is required to produce evaluated analytical data.

[4] Amsden, D.M. and R. T. Amsden, Eds., "QC Circles: Applications, Tools, and Theory," American Society for Quality Control, Milwaukee, WI 53203 (1976).

A collection of 23 papers on various aspects of quality circles, subdivided into the following groups: QC circle applications, USA and Japan; behavioral theory and the QA circle; QC circle tools. The papers were first presented at a conference sponsored by ASQC in Toronto in 1978.

[5] Anonymous, "Quality Assurance of Environmental Measurements," Information Transfer, Inc., 9300 Columbia Blvd., Silver Spring, MD 20910.

A collection of 33 papers presented at a national conference held in Denver, CO, November 27–29, 1978. The papers are classified as follows: legal and institutional problems; quality assurance methodology; water sampling and measurements; air sampling and monitoring; wastewater analysis; laboratory method control.

[6] American National Standards Institute, "ANSI - ASQC Standard A3–1978 Quality Systems Terminology," American National Standards Institute, 1430 Broadway, New York, NY 10018.

A collection of definitions of terms used in quality control and related activities.

[7] American Society for Testing and Materials, "ASTM Manual on Presentation of Data and Control Chart Analysis," STP 15D, ASTM, Philadelphia, PA 19103.

This is an excellent source of information on ways to present data by means of charts and statistical representation. The basic concepts of several kinds of Shewhart control charts are discussed with examples. Tables are provided to calculate control limits.

[8] ASTM C1009 "Establishing a Quality Assurance Program for Analytical Chemistry Laboratories Within the Nuclear Industry," ASTM, Philadelphia, PA 19103.

This provides guidance for establishing a quality assurance program within the nuclear industry. Major topics include: organization; QA programs; training and qualifications; procedures; laboratory records; control of records; control of equipment and materials; control of measurements; deficiencies and corrective actions.

[9] ASTM D2777 "Standard Practice for Determination of the Precision and Bias of Methods of Committee D19 on Water," ASTM, Philadelphia, PA 19103.

Detailed procedure, including statistical evaluation of data, for collaborative tests designed to evaluate analytical methodology.

[10] ASTM D3614 "Evaluating Laboratories Engaged in Sampling and Analysis of Atmospheres and Emissions," ASTM, Philadelphia, PA 19103.

A detailed set of criteria which represent minimum performance standards and hence are useful for evaluating laboratories engaged in air and emissions measurements.

[11] ASTM D3856 "Evaluating Laboratories Engaged in Sampling and Analysis of Water and Waste Waters," ASTM, Philadelphia, PA 19103.

A discussion of minimum requirements for laboratories engaged in water analysis of various kinds and hence useful in evaluation of such laboratories.

[12] ASTM D4210 "Intralaboratory Quality Control Procedures and a Discussion on Reporting Low-level Data," ASTM, Philadelphia, PA 19103.

The various options for reporting data at or below detection levels are presented.

[13] ASTM E173 "Conducting Interlaboratory Studies of Methods of Chemical Analysis of Metals," ASTM, Philadelphia, PA 19103.

This standard describes procedures for sample distribution and interlaboratory testing to determine repeatability and reproducibility of methods for the chemical analysis of metals. Both graphical and numerical methods are described for analyzing the test data.

[14] ASTM E177 "Standard Recommended Practice for Use of the Terms Precision and Accuracy as Applied to Measurement of a Property of a Material," ASTM, Philadelphia, PA 19103.

This document outlines some general concepts regarding the terms "precision" and "accuracy." Notation for use in statements of precision and accuracy of measurement methodology is recommended. Statements of precision should always be used with descriptors to indicate causes. Accuracy depends on both precision and bias, and each component of inaccuracy needs to be specified when describing the performance of methodology.

[15] ASTM E178 "Standard Recommended Practice for Dealing with Outlying Observations," ASTM, Philadelphia, PA 19103.

A general discussion of the statistical criteria for identification and rejection of outlying data. The text contains examples and tables useful in making statistical judgments.

[16] ASTM E305-83 "Standard Practice for Establishing and Controlling Spectrochemical Analytical Curves," ASTM, Philadelphia, PA 19103.

Systematic and random errors that occur in obtaining data are reviewed. Background corrections are considered for linear and logarithmic readout systems. Calibration procedures are given, including the reference materials to be used and the generation of data. Procedures are provided for constructing the analytical curve, fitting the regression curve, and evaluating curve fit. Control of curve shift and notation is described.

[17] ASTM E548 "Standard Recommended Practice for Generic Criteria for Use in Evaluation of Testing and/or Inspection Agencies," ASTM, Philadelphia, PA 19103.

This practice provides a guide for the information needed to evaluate the technical competence of a laboratory. It may be used by accreditors or by laboratories themselves for self-appraisal.

[18] ASTM E691 "Practice for Conducting an Interlaboratory Test Program to Determine the Precision of Test Methods," ASTM, Philadelphia, PA 19103.

This is the basic practice developed by ASTM Committee E-11 on Statistical Methods to provide guidance for the conduct of collaborative testing exercises. In addition, it contains extensive and detailed diagnostic procedures (tabular, graphic, and statistical) for investigating variability in test data.

[19] ASTM E748 "Quality Assurance Procedure for Spectrographic Laboratories," ASTM, Philadelphia, PA 19103.

This guide outlines the requirements for a system of quality assurance for the spectrochemical laboratory. It does not include a detailed description for setting up such a program.

[20] ASTM E882 "Accountability and Quality Control in the Chemical Analysis Laboratory," ASTM, Philadelphia, PA 19103.

This standard guide prepared by ASTM Committee E-3 on Chemical Analysis of Metals describes approaches to be used by management in establishing a system of documented records to assure users of services that a specified level of precision is achieved in routine performance and that the data reported were obtained from the samples submitted.

[21] Barnard, Jr., A.J., R. M. Mitchell, and G. E. Wolf, "Good Analytical Practices in Quality Control," *Anal. Chem.* 50 (No. 12): 1079A-86A (1978).

This paper emphasizes that a laboratory must utilize good laboratory practices as a prerequisite for the more formal aspects of a QA program.

[22] Belanger, B.C., "Traceability: An Evolving Concept," *ASTM Standardization News*, 8: 22, ASTM, Philadelphia, PA 19103 (1980).

A general discussion of the traceability concept, mostly from a calibration point of view.

[23] Belanger, B.C., "Measurement Assurance Programs Part I: General Introduction," NBS SP 667-I, National Bureau of Standards, Gaithersburg, MD 20899 (1984).

A general discussion of the NBS MAP program. *See also* Reference 39.

[24] Berman, G.A., Ed., "Testing Laboratory Performance: Evaluation and Accreditation," NBS SP 591, NBS, Gaithersburg, MD 20899.

This book contains the texts of 60 papers presented at a National Conference on Testing Laboratory Performance Evaluation and Accreditation in 1979. The areas covered include: overview of evaluation and accreditation; evaluation technology; health services accreditation programs; accreditation systems and concepts; quality control; evaluation programs and systems; international coordination.

[25] Bervere, J.M. and I. D. Macaulay, "Data are for Looking At or Quality Control Through Interpretation," In: Water Quality Parameters, pp. 550–562, ASTM STP 573, ASTM, Philadelphia, PA 19103.

A logical interpretation procedure is described to eliminate errors, detect and evaluate anomalies, and assure that data are valid.

[26] Besterfield, D.H., *Quality Control*, Prentice-Hall (1979).

A textbook covering: introduction to quality; fundamentals of statistics; control charts for variables; fundamentals of probability acceptance sampling; product liability; computer and quality control. Industrial slant.

[27] Bittner, J., et al., (Committee Report) "Provisional Recommendations on Quality Control in Clinical Chemistry, Part 1: General Principles and Terminology," *Clin. Chem.* 22 (No. 4): 532–540 (1978).

A good discussion of terminology and basic principles of quality control as related to clinical chemical measurements.

[28] Bordner, R.H., "Quality Assurance for Microbiological Analyses of Water," In: Quality Assurance for Environmental Measurements, ASTM

STP 867, J. K. Taylor and T. W. Stanley, Eds., American Society for Testing and Materials, Philadelphia PA 19103, pp. 133–143 (1985).

The establishment of microbiological contaminant limits for water and wastewater, and the implementation of enforcement programs which require monitoring of these limits have emphasized the need to develop and practice quality assurance programs. A broad scope QA program is essential for microbiological determinations. The basic elements of an intralaboratory quality control program are described, which include controls over personnel, facilities, sampling techniques, laboratory operations, analytical procedures, and data reporting. Formal written analytical procedures, a QA protocol, and complete QC records are indispensable to a QA program. The laboratory should participate in an interlaboratory QC program established to maintain minimal analytical and operational standards. The U.S. Environmental Protection Agency (EPA) certification program for analyzing public drinking water supplies is described as an example of a national interlaboratory program. Guidelines provided by the EPA for laboratory QA programs are outlined.

[29] Bordner, R. and J. Winter, Eds., "Microbiological Methods for Monitoring the Environment," EPA-600/8-78–017 (1978).

Discussed are: analytical operations; general operations; analytical methodology; quality control; laboratory management. Water is the environment of concern.

[30] Brossman, M.W., T. J. Hoogheem, and R. C. Splinter, "Quality Assurance Project Plans—A Key to Effective Cooperative Monitoring Programs," In: Quality Assurance for Environmental Measurements, ASTM STP 867, J. K. Taylor and T. W. Stanley, Eds., American Society for Testing and Materials, Philadelphia, PA 19103, pp. 53–61 (1985).

The use of cooperative monitoring programs, involving the co-mingling of data of the regulated community and the "regulators," can lead to effective utilization of resources and contribute to cooperative solutions for our pollution control programs. Such benefits can only be realized, however, under controls which ensure the comparability and validity of data obtained through well-defined roles and work plans. A key to this control is the Work/Quality Assurance (QA) Project Plan developed cooperatively by the type of team reflected in the authorship of this paper—namely, by federal, industry, state, and municipal representatives. A comprehensive Work/Quality Assurance Project/Work Plan Guidance document has been developed in the Office of Water, U.S. Environmental Protection Agency (EPA), which should effectively meet QA requirements for a wide range of users and types of water monitoring tasks. This guidance document is designed to cover all aspects of water environmental measurement from field sampling through laboratory analysis and data reduction and, where applicable, computer input. The document integrates a work plan with a QA plan, eliminating the dual effort of overlapping plan developments and assuring practical incorporation of

quality assurance and quality control. In addition, the combination of the work plan and QA plan provides a comprehensive basis for establishing sound coordination and agreement for cooperative monitoring tasks.
This guidance document is currently in pilot implementation throughout the United States.

[31] Bryson, J., et al., "Bibliography on Laboratory Accreditation," NBSIR 82-2523, NBS, Gaithersburg, MD 20899 (1982).

This bibliography is a compilation of annotated references of published information on laboratory accreditation programs. An index facilitates search on the basis of country, subject area, and date of publication.

[32] Cadle, S.H., "Internal Consistency Checks as Part of a Data Validation Process for Chemical Constituents in Ambient Particulate Samples," In: Quality Assurance for Environmental Measurements, ASTM STP 867, J. K. Taylor and T. W. Stanley, Eds., American Society for Testing and Materials, Philadelphia, PA 19103, pp. 89-99 (1985).

Internal consistency checks are an excellent means of validating data obtained during field studies of ambient particulate. The specific method used depends on the purpose and scope of the program. Techniques discussed include inspection of data for outliers, intersampler comparison, and ion balance and mass balance techniques. The use of these methods is illustrated with data collected in a variety of field programs. It is concluded that internal consistency checks should be incorporated into the quality assurance program of all field studies.

[33] Cali, J.P., "Problems of Standardization in Clinical Chemistry," *Bull. World Health Org.* 48: 721-6 (1973).

If analytical results in clinical chemistry are to be made meaningful (i.e., accurate, precise, and specific) a systematic approach to their attainment is necessary. Furthermore, because this system is so complex in scope and the need for it is so widespread, it will require international coordination. Agreement on the units of measurement, the production and certification of standard reference materials, and the development of reference methods of demonstrated accuracy will require the support of all segments of clinical chemistry.

[34] Cali, J.P., et al., "The Role of Standard Reference Materials in Measurement Systems," NBS Monograph 148, NBS, Gaithersburg, MD 20899 (1975).

This publication is a guide to the use of Standard Reference Materials (SRMs) and should be useful to all users of SRMs, particularly those in countries developing national measurement systems. It is not intended to be an exhaustive description of the NBS-SRM program, but rather a review of the role SRMs play in the measurement system, how SRMs are certified, and what the certification means. To illustrate the use of SRMs, several selected industries are described in which SRMs have made significant contributions.

[35] Campbell, S. and H. Scott, "Quality Assurance in Acid Precipitation Measurements," In: Quality Assurance for Environmental Measurements, ASTM STP 867, J. K. Taylor and T. W. Stanley, Eds., American Society for Testing and Materials, Philadelphia, PA 19103, pp. 272–283 (1985).

The growing interest in acid deposition has lead to a proliferation of laboratories engaged in such studies. High-level quality assurance (QA) procedures are required for each program to standardize the diverse measurement methods in use and to determine the validity of differences in measurements widely separated in space and time. Both in-laboratory (quality control) and external (quality assurance) procedures are required. A complete QA program for acid precipitation measurements must address: program objectives; site selection and operation; operator selection and training; sample collection, handling, and analyses; and data checking, storage, retrieval, and transmission. Objective criteria must be developed for detecting adulterated samples and invalid data. Appropriate laboratory and field blanks must be collected and analyzed. Standard techniques (sample spiking, replicate analysis of standards and samples) should ensure the reliability of analytical results. Relevant quality assurance data, including analytical detection limits, blank values, and the variability of replicate determinations, must be supplied with each data transmittal.

[36] CFR Title 21, Food and Drugs, Chapter 1, FDA, D Part 38, "GLPs for Non-Clinical Laboratory Studies."

This work provides federal quality assurance requirements for this subject.

[37] CFR Title 40, Part 792, "Toxic Substances Control", *Federal Register* 48(No. 230), November 28, 1983, p. 53922.

This work provides federal quality assurance requirements for this subject.

[38] CFR Title 42, Public Health, Chapter 1, Public Health Service, Part 74, "Clinical Laboratories."

This work provides federal quality assurance requirements for this subject.

[39] Croarkin, C., "Measurement Assurance Programs Part II: Development and Interpretation," NBS SP 676-II (1984).

This companion to Part I (*see* Ref. 23) provides: discussion of development of MAPs; discussion of measurement error; and use of a check standard; and interpretation of special MAPs.

[40] Crosby, P.B., *Quality is Free*, McGraw Hill Book Co. (1979).

A philosophical book on "how to manage quality—so that it becomes a source of profit for your business." No quantitative discussion is given. Real-life case histories are sprinkled throughout the book.

[41] Crummett, W.B. and J. K. Taylor, "Guidelines for Data Acquisition and Data Quality Evaluators in Environmental Chemistry," IUPAC Collaborative Interlaboratory Studies in Chemical Analysis, H. Egan and T. S. West, Eds., Pergamon Press, Oxford and New York (1982).

This paper is a review and critique of the "Guidelines For Data Acquisition and Data Quality Evaluation In Environmental Chemistry." The components of analytical chemistry are identified as planning, quality assurance, sampling, calibration and standardization, measurement, validation, and documentation. The guidelines recognize that a measure of uncertainty is associated with each of these components. Some of the key features of each of these components are discussed, and the conclusion is drawn that for accurate environmental chemical analysis, all these parameters must be carefully taken into account.

[42] Currie, L.A., "Limits for Qualitative Detection and Quantitive Determination," *Anal. Chem.* 40: 586–93 (1968).

The occurrence in the literature of numerous, inconsistent, and limited definitions of a detection limit has led to a reexamination of the questions of signal detection and signal extraction in analytical chemistry and nuclear chemistry. Two limiting levels have been defined: LD—the "true" net signal level which may be a priori expected to lead to detection; and LQ—the level at which the measurement precision will be satisfactory for quantitative determination. Exact defining equations as well as series of working formulas are presented both for the general analytical case and for radioactivity. The latter, assumed to be governed by the Poisson distribution, is treated in such a manner that accurate limits may be derived for both short- and long-lived radionuclides, either in the presence or absence of interference. The principles are illustrated by simple examples of spectrophotometry and radioactivity, and by a more complicated example of activation analysis in which a choice must be made between alternative nuclear reactions.

[43] Deming, W.E., "On Some Statistical Aids Toward Economic Production," *Interface* 5(No. 4): 1–15 (August 1975).

This work describes management's responsibility for quality outputs from a statistical point of view.

[44] Dewar, D.L., "Quality Circles—Answers to 100 Frequently Asked Questions," Quality Circle Institute, P.O. Box Q, Red Bluff, CA 96080 (1979).

An illustrated question and answer booklet describing the objectives, organization, and operation of quality circles.

[44A] Dorko, W.D. and E. E. Hughes, "Special Calibration Systems for Reactive Gases and Other Difficult Measurements," In: Calibration and Sampling for Atmospheric Measurements, ASTM STP 957, J. K. Taylor, Ed., American Society for Testing and Materials, Philadelphia, PA 19103 (1987).

Procedures to prepare reactive gas standards are described. Ways to test for stability of mixtures are discussed.

[45] Eggenberger, L.M., "Establishing an Environmental Quality Assurance Program," In: Quality Assurance for Environmental Measurements, ASTM STP 867, J. K. Taylor and T. W. Stanley, Eds., American Society for Testing and Materials, Philadelphia, PA 19103, pp. 391–406 (1985).

Specific steps leading to the establishment of an environmental quality assurance program in an analytical laboratory are enumerated and discussed. Aspects of a model quality assurance program that applies fundamental principles and methods to selected laboratory instrumentation, and analytical protocols for the attainment of accurate, reliable results are reviewed. Discussion includes specific factors, such as method validation, certification, sample handling, security, documentation, and continuing data assessment. Objectives and problem areas characteristic of newly implemented programs, such as building a data base, interim control limits, reference standards, and categorizing and prioritizing prospective control areas, are included in the discussion.

[46] Eisenhart, C., "Realistic Evaluation of the Precision and Accuracy of Instrument Calibration Systems," Reprinted in Ref. 81.

A discussion of the statistical basis for specification of the precision of a measurement process. The need for statistical control is emphasized. Ways to express precision and accuracy are discussed.

[47] Environmental Protection Agency, "EPA Handbook for Analytical Quality Control in Water and Waste Water Laboratories," EPA 600/4–79–019 (March 1979).

One of the fundamental responsibilities of water and wastewater management is the establishment of continuing programs to ensure the reliability and validity of analytical laboratory and field data gathered in water treatment and wastewater pollution control activities. This handbook is addressed to laboratory directors, leaders of field investigations, and other personnel who bear responsibility for water and wastewater data. Subject matter of the handbook is concerned primarily with quality control for chemical and biological tests and measurements. Chapters are also included on QC aspects of sampling, microbiology, biology, radiochemistry, and safety as they relate to water and wastewater pollution control. Sufficient information is offered to allow the reader to inaugurate or reinforce programs of analytical QC that emphasize early recognition, prevention, and correction of factors leading to breakdowns in the validity of water and wastewater pollution control data.

[48] EPA, "EPA Quality Assurance Handbook for Air Pollution Measurement Systems," Vol. I, Principles, EPA-600/9–76–005; Vol. II, Ambient Air Specific Methods, EPA 600/4–77–027a; Vol. III, Stationary Source Specific Methods, EPA 600/4–77–027b; U.S. EPA-ORD, EMSL Research Triangle Park, NC 27711.

Volume I contains brief discussions of the elements of quality assurance. Expanded discussions of technical points and sample calculations are included in Appendices. Volume II contains guidelines for QA of ambient air measurement systems, while Volume III contains similar information for source measurements.

[49] EPA QAMS-005/80, "Interim Guidelines and Specifications for Preparing Quality Assurance Project Plans."

The U.S. EPA requires that 14 points be addressed by contractors and the agency itself when developing QA plans for environmental measurement projects and programs. This document explains the rationale for this policy and gives specific guidance for preparing acceptable project plans.

[50] Erdmann, D.E. and J. D. Thomas, "Quality Assurance of U.S. Geological Survey Water-Quality Field Measurements," In: Quality Assurance for Environmental Measurements, ASTM STP 867, J. K. Taylor and T. W. Stanley, Eds., American Society for Testing and Materials, Philadelphia, PA 19103, pp. 110–115 (1985).

Each year many thousands of water quality measurements are made in the field by personnel of the U.S. Geological Survey's Water Resources Division. To ensure the accuracy of these measurements, an external quality assurance program was officially established in March 1979. The initial phase of this program was concerned with specific conductance and pH measurements. Reference samples are submitted semiannually to field analysts for measurement of these parameters with the same techniques and instruments used in the field. Both the personnel and the instruments involved in making the determinations are recorded. When the data are complete, a report defining the quality of the analytical results is prepared and circulated to appropriate district, regional, and national offices. Field data, in addition to their immediate use, are normally placed in a national data base and are available to all hydrologists. As a result, the quality of field measurements can have a profound effect on their investigations. The quality assurance program described is designed to ensure that the field data of the Water Resources Division are highly accurate.

[51] Filliben, J.J., "Testing Basic Assumptions in the Measurement Process," In: Validation of the Measurement Process, J. R. DeVoe, Ed., ACS Symposium Series No. 63, American Chemical Society, Washington, DC 20036 (1977).

The basic assumptions are: randomness, fixed location, fixed variation, and fixed distribution. Various statistical tests for testing these are described.

[52] Friedman, L.C. and D. E. Erdmann, "Quality Assurance Practices for the Chemical and Biological Analysis of Water and Fluvial Sediments," Book 5-Chapter A6, In: Techniques of Water Resource Investigations of the USGS, U. S. Government Printing Office, Washington, DC 20402 (1982).

This is part of an extensive publication of the U.S. Geological Survey on methods for water resources investigations. It relates to the QA aspects of such investigations and deals with such matters as: evaluation of precision and bias of methods; interferences; quality assurance practices; preparation of reference samples; and quality control as related to inorganic, organic, and biological measurements.

[53] Gaft, S. and F. D. Richards, "Quality Assurance at Ford Motor Co., Central Laboratory—A Dynamic Approach to Laboratory Quality," In: ASTM STP 814, ASTM, Philadelphia, PA 19103.

A system of internal audits that a laboratory may use to evaluate its quality assurance program and to assess the quality of its outputs.

[54] Garfield, F.M., "Quality Assurance Principles for Analytical Laboratories," AOAC, Arlington, VA 22209 (1984).

A discussion of quality assurance from a management point of view.

[55] Glaser, J.A., et al., "Trace Analysis for Waste Waters—Method Detection Limit," *Environ. Sci. Technol.* 15: 1426–35 (1981).

A procedure is described whereby the MDL (limit of detection for a complete methodology) may be evaluated. This is based upon the standard deviation at lowest level of measurement, s_o, and is defined as MDL = $3s_o$.

[56] Hainline, Jr., A., "Quality Performance Before Quality Control," In: Laboratory Management, p. 27 (October 1974).

The author stresses the need for good laboratory practices before a QC program is effective. Seven reasons—management related—are given why QC systems fail.

[57] Handy, R.W., H. L. Crist, and T. W. Stanley, "Quality Assurance for Personal Exposure Monitoring," In: Quality Assurance for Environmental Measurements, ASTM STP 867, J. K. Taylor and T. W. Stanley, Eds., American Society for Testing and Materials, Philadelphia, PA 19103, pp. 284–286 (1985).

A personal exposure monitoring program has been carried out to field-test methodologies for sample collection and analysis. The basic study involved the collection of personal air, breath, and water samples. Quality assurance procedures were implemented during the sampling period and included exposing field controls and blanks and the collection of duplicates, some of which were analyzed by a reference laboratory. Daily checks on instrument performance and calibration were performed, and precision quality control charts were maintained. The analysis of performance audit samples was a major component of the quality assurance protocol. These results clearly show that background on the sampling devices adversely affects data quality and must be monitored closely.

[58] Hendrix, C.D., "What Every Technologist Should Know About Experimental Design," *CHEMTECH*, p. 167 (March 1979).

The total experimental effort required in process and product development work can be greatly reduced through the use of two-level factorial designs. This is true because a large number of variables can be thoroughly studied for little more than the cost of studying just one variable. The effect of each and every variable will be determined by comparing averages against averages so that the impact of random errors is minimized. Statistically designed experiments were used in agricultural field trials more than fifty years ago, and have been used in industrial experimentation for twenty years. More sophisticated designs and methods of analysis have appeared since digital computers became readily available in the mid-1960s. But the most useful and most widely used designs are the ones that are the easiest to practice. That is the two-level factorial series.

[59] Henry, J.A. and L. A. Knowler, "Two-Day Intensive Training Course in Elementary Statistical Quality Control," 3rd ed., ASQC, Chicago Section, Chicago, IL (1980).

This work consists of twelve chapters, mostly on control charts for acceptance sampling. This is the 70-page manual used in a two-day course offered by ASQC.

[60] Hertz, H.S. and S. N. Chesler, Eds., "Trace Organic Analysis: A New Frontier in Analytical Chemistry," NBS SP 519, National Bureau of Standards, Gaithersburg, MD 20899 (1979).

Researchers in diverse areas must currently perform critical analyses on minute quantities of organic compounds in various matrices. It was the aim of this symposium to bring together these scientists to discuss their common problems and to explore current and impending technology for organic analyses. Emphasis was placed on the total analysis, from collecting the sample through interpreting the results, rather than upon the measurement only. The proceedings consist of a series of 12 invited papers by experts and particulary appropriate contributed papers. Topics covered in the proceedings are as follows: sampling and sample handling for trace organic analysis, state-of-the-art analytical systems, analytical techniques on the horizon, analysis of nutrients, analysis of organic pollutants and their metabolites in the ecosystem, analysis of drugs in body fluid, analysis of food toxicants, and analysis of hormones and neurotransmitters.

[61] Hillman, J.J., "Certification of Local Laboratories Analyzing Drinking Water in Nonprimacy States," In: Quality Assurance for Environmental Measurements, ASTM STP 867, J. K. Taylor and T. W. Stanley, Eds., American Society for Testing and Materials, Philadelphia, PA 19103, pp. 103–109 (1985).

This paper discusses the U.S. Environmental Protection Agency (EPA) certification procedures for certifying local laboratories analyzing drinking water for compliance with the Safe Drinking Water Act. A local laboratory is defined as a nonprincipal state laboratory and may include state, county, municipal, federal, or commercial laboratories. The EPA certification program has recently been updated and clarified in the 1983 edition of the EPA certification manual. Critical elements for certification in chemistry, microbiology, and radiochemistry have been revised and include the preparation of a quality assurance (QA) plan. Requirements to maintain certification as well as criteria and procedures for downgrading a laboratory's certification status are described in the manual. Technical services to be provided by the EPA have been expanded. The paper briefly describes the certification program, changes that are incorporated in the new manual that will affect the certification of local laboratories, and potential future changes.

[62] Hubaux, A. and G. Vos, "Decision and Detection Limits for Linear Calibration Curves," *Anal. Chem.* 42: 849–55 (1970).

For linear calibration curves, two kinds of lower limits may be connected to the notion of confidence limits—a decision limit, the lowest signal that can be distinguished from the background, and a detection limit, the content under which, a priori, any sample may erroneously be taken for the blank. From a few algebraical and computational developments, several practical rules are deduced to lower these limits. The influence of the precision of the analytical method, the number of standards, the range of their contents, the various modes of their repartition, and the replication of measurements on the unknown sample are studied from a statistical point of view.

[63] Hughes, E.E. and J. Mandel, "A Procedure for Establishing Traceability of Gas Mixtures to Certain NBS SRMs," EPA/781- 0101 (1981).

A protocol is given that defines how "traceable" certified reference materials are to be prepared and intercompared with SRMs of essentially the same composition. The procedure the EPA uses to verify the claims of vendors is described.

[64] Hunter, J. Stuart, "Calibration and the Straight Line: Current Statistical Practices," *AOAC Journal* 64: 574 (1982).

The use of linear regression to fit a line to calibration data is described. The statistical uncertainty of the line so fitted is discussed.

[65] ILAC Task Force E, "Guidelines for the Determination of Recalibration Intervals of Testing Equipment Used in Testing Laboratories," International Organization for Legal Metrology (OIML). International Document No. 10, OIML, 11 Rue Twigot, Paris 95009, France (1984).

Both intuitive and statistical guidelines are discussed for establishing realistic calibration schedules for measuring equipment. While slanted toward physical measurements, the guidelines should be useful in decisions on similar matters related to chemical measuring equipment.

[66] Ingamells, C.O. and P. Switzer, "A Proposed Sampling Constant for Use in Geochemical Analysis," *Talanta* 20: 547–67 (1973); 23: 263 (1974).

The problem of sampling for the analysis of heterogeneous materials is discussed. When the sample size is decreased, as in the case of subsampling, the standard deviation of sample composition increases. A sampling constant, K_s, is proposed which relates weight of the analytical subsample, W, to the relative standard deviation, R%, by the expression $K_s = R^2W$.

[67] Inhorn, S.L., Ed., "Quality Assurance Practices for Health Laboratories," American Public Health Association, Washington, DC (1978).

The 20 multiauthored chapters in this book deal with various aspects of quality assurance for laboratories engaged in health-related (largely clinical) measurements. The first 5 chapters are general and are concerned with: laboratory management; certification; facilities and services; measurement evaluation; and data transmission. The remaining chapters discuss specific quality assurance practices, largely in chemical measurement areas.

[68] Ishikawa, Kaoru, "Guide to Quality Control," Asian Production Organization, Tokyo, Japan (1976).

A guide to quality control in industry. Excellent discussion of: how to collect data; histograms; cause-effect diagrams; pareto analysis; control charts; sampling; and examples and problems.

[69] ISO Guide 25, "Guidelines for Assessing the Technical Competence of Testing Laboratories," American National Standards Institute, 1430 Broadway, New York, NY 10018.

An international standard, similar to ASTM E-548, of criteria considered to be essential for quality outputs.

[70] ISO Guide 30, "Terms and Definitions Used in Connection with Reference Materials," American National Standards Institute, 1430 Broadway, New York, NY 10018.

Terms used in describing, measuring, certifying, and reporting reference materials are defined.

[71] ISO Guide 31, "Contents of Certificates of Reference Materials," American National Standards Institute, 1430 Broadway, New York, NY 10018.

The informational content of certificates issued for reference materials is specified and described in considerable detail.

[72] Juran, J.M., *Quality Control Handbook,* 3rd ed., McGraw Hill Book Co. (1974).

This work is a compilation of 48 articles slanted mostly to QC of manufacturing and manufactured products.

[73] Kanare, H.W., *Writing the Laboratory Notebook,* American Chemical Society, Washington, DC 20036 (1985).

This book reviews the principles of proper scientific notekeeping. It goes beyond the mechanics of filling in the pages of a notebook and provides specific recommendations to help develop a flexible style of notekeeping that will serve in a variety of laboratory situations. Examples of notebook format and entry are included, and summaries of current notekeeping practices in some universities, industrial laboratories, and government research centers are presented. Alternatives to handwritten bound notebooks are discussed, and comparison is made between the electronic notebook and the handwritten notebook. The hardware of notekeeping is described. Pens, paper, ink, notebook binding, and proper storage are all detailed.

[74] Kateman, G. and F. W. Pijkers, *Quality Control in Analytical Chemistry,* J. Wiley & Sons, New York, NY (1981).

Discusses topics such as sampling, analysis, data processing, and organization, largely from a systems operation point of view.

[75] Kieffer, L.J., Ed., "Calibration and Related Measurement Services of the National Bureau of Standards," NBS SP250, National Bureau of Standards, Gaithersburg, MD 20899 (1982).

This publication provides descriptions of the currently available NBS calibration services, special test services, and measurement assurance programs. In addition, each section describing specific services contains references to additional publications giving more detail about the measurement techniques and procedures used. This revised edition reflects the services available as of the first quarter of 1982. NBS Special Publication 250 was last issued in 1980. The Appendix to SP250 is reissued every 6 months (April and October). It lists current prices for the services described in this publication and the NBS points of contact (addresses and phone numbers) from whom additional information can be obtained.

[76] Kirchmer, C.J., "Quality Control in Water Analysis," *Environ. Sci. Technol.* 17: 174A-181A (1983).

A general discussion of quality control as it is related to water analysis. Six sources of bias of measurements are discussed together with ways to minimize them. Ways to achieve interlaboratory quality control for measurement programs are considered.

[77] Kratochvil, B.G. and J. K. Taylor, "Sampling for Chemical Analysis," *Anal. Chem.* 53: 924A (1981).

This work provides a general discussion of sampling with special emphasis on bulk materials.

[78] Kratochvil, B.G. and J. K. Taylor, "A Survey of the Recent Sampling Literature on Sampling for Chemical Analysis," NBS Tech Note 1153, National Bureau of Standards, Gaithersburg, MD 20899 (January 1982).

This work provides an annotated listing containing references to various publications on sampling.

[79] Kratochvil, B.G., D. Wallace, and J. K. Taylor, "Sampling for Chemical Analysis," *Anal. Chem. Reviews* 56: 113R (1984).

The sampling literature for the period 1975 to 1984 is reviewed and summarized in the following areas of chemical analysis: general considerations; theory; standards; mineralogy; soils; metallurgy; oil and gas; water; sediments; atmosphere; biology, agriculture and food; and clinical and medical. The review is based on 541 papers that are cited in the text.

[80] Kropp, E.L. and D. N. Flannery, "Guilty or Not Guilty – Only Your Statistician Knows for Sure," *Environmental Forum*, p. 23 (April 1984).

"The measurement errors, inherent in most analytical processes, are particularly troublesome when the sample result is outside of compliance limits but within probable error margin." A plea for better measurements and their sane interpretation is given.

[81] Ku, H.H., Ed., "Precision Measurement and Calibration: Statistical Concepts and Procedures," NBS SP 300, Vol. 1, National Bureau of Standards, Gaithersburg, MD 20899.

A compendium of 44 papers dealing with the statistical aspects of calibration and precision measurement.

[82] Ku, H.H., "Statistical Concepts in Metrology, in Precision Measurement and Calibration," [Ref. 81] pp. 296–328.

A general discussion of the statistical relations most often used to interpret the quality of measurement data.

[83] LaFleur, P.D., Ed., "Accuracy in Trace Analysis, Sampling, and Sample Handling," NBS SP 422, National Bureau of Standards, Gaithersburg, MD 20899 (1976).

This book is the formal report of the proceedings of the 7th Materials Research Symposium: Accuracy in Trace Analysis. This volume contains the 98 invited and contributed papers presented at the symposium which treat the general question of accuracy and the problems of sampling and sample handling, as well as the usually discussed analytical methodology. Many important techniques and methods are described, and extensive references are presented to give deeper insight into the problems of obtaining accurate results in trace analytical chemistry. Accordingly, this volume should not only stimulate greater interest in research in these areas but should provide a valuable guide for everyday analytical problems.

[84] Linnig, F.J. and J. Mandel, "Which Measure of Precision? The Evaluation of the Precision of Analytical Methods Involving Linear Calibration Curves," *Anal. Chem.* 36(No. 13): 25A-32A (December 1964).

Ways to evaluate the precision of a method as contrasted to measurement data are discussed from the standpoints of: replication error; scatter about calibration line; and uncertainty of the calibration line.

[85] Locke, J.W., Ed., "Laboratory Accreditation—Future Directions in the United States," NBS Special Publication 632, National Bureau of Standards, Gaithersburg, MD 20899 (March 1982).

This is a reprint of the proceedings of a workshop held at NBS in November 1981 to discuss various approaches to laboratory accreditation. It contains papers presented by various persons discussing the need for accreditation, modes of accreditation, and ways to evaluate accreditors. Various international accreditation systems are described as well as the general approaches used in the USA.

[86] Lott, J.A., "Laboratory Personnel: The Most Important Aspect of Quality Control," *Medical Instrumentation* 8: 22–25 (1974).

The quality of service provided by laboratory personnel who handle specimens or provide laboratory information depends completely on staff motivation and performance. Some training of new personnel can be done in the laboratory, but laboratories are becoming increasingly dependent on trained staff. One of the biggest problems in laboratories is human error. Laboratory instruments should be designed to minimize stress and boredom. Much more redundancy and computer checking of laboratory results is needed to reduce the frequency of laboratory error.

[87] Mandel, J., "Fitting Straight Lines When Both Variables are Subject to Error," *J. Qual. Technol.* 16 (No. 1): 1–14 (1984).

Least squares linear regression is one of the most widely used statistical techniques. Almost all textbooks or statistical methods provide the necessary formulas for the fitting process, based on the assumption that there is no error in the independent variable. How these formulas should be modified when both variables are subject to error is dealt with in detail using as an example an interlaboratory study.

[88] Mandel, J. and F. J. Linnig, "Study of Accuracy in Chemical Analysis Using Linear Calibration Curves," *Anal. Chem.* 29: 743–9 (1957).

In situations characterized by linear calibration curves, such as the relation between "found" and "added" in studies of accuracy in chemical analysis, the usual method for deriving confidence intervals for the slope and the intercept of the fitted straight line may lead to erroneous conclusions. The difficulty results from the interdependence of multiple conclusions drawn from the same data, especially when there is a strong correlation between the parameters involved. The method of joint confidence regions eliminates these difficulties and has the further advantage of allowing for the evaluation of the uncertainty of the calibration line as a whole, as well as of any values or functions of values derived from it.

[89] Mandel, J. and T. Lashoff, "Interpretation and Generalization of Youden's Two-Sample Diagram," *J. Qual. Tech.* 6: 22–36 (1974).

This work gives a detailed statistical treatment of Youden's two-sample approach for identification of precision and bias.

[90] Margolis, S.A., Ed., "Reference Materials for Organic Nutrient Measurement," NBS SP 635 National Bureau of Standards, Gaithersburg, MD 20899 (1982).

Proceedings of a workshop held at NBS October 23, 1980. Contains some of the papers presented plus reports of group discussion workshops on water soluble vitamins, carbohydrates, fat soluble vitamins, cholesterol, and fat.

[91] Mavrodineanu, R., J. I. Shultz, and O. Menis, Eds., "Accuracy in Spectrophotometry and Luminescent Methods," NBS SP 378, National Bureau of Standards, Gaithersburg, MD 20899.

Proceedings of a conference held at NBS March 22- 24, 1972. A collection of 18 papers discussing accuracy problems and standards for accurate measurements in spectrophotometry and luminescence.

[92] Mavrodineanu, R., Ed., "Procedures Used at NBS to Determine Selected Trace Elements in Biological and Botanical Materials," NBS SP 492, National Bureau of Standards (1977).

Methods fall under the following groups: sample preparation; neutron activation analysis; spark source mass spectrometry isotope dilution; atomic absorption and flame emission; molecular spectrophotometry; and polarography. A total of 13 papers are included.

[93] McNish, A.G., "Dimensions, Units, and Standards," *Physics Today* 10: 19–25 (1957).

A general discussion of the basis for physical measurements, i.e., the units and standards developed to make measurements consistent.

[94] McNish, A.G., "Fundamentals of Measurement," Electrotechnology, Science & Engineering Series 53 (1983).

A basic discussion of measurement and the propagation of error, as related to the fundamental physical quantities but also good for general education in metrology.

[95] Meinke, W.W., Ed., "Trace Characterization, Chemical and Physical," NBS Monograph 100, National Bureau of Standards, Gaithersburg, MD 20899.

A compilation of papers presented at a symposium of the same name held at NBS in October 1966.

[96] MIL-STD-105, "Sampling Procedures and Tables for Inspection by Attributes," Government Printing Office, Washington, DC 20402.

This publication contains sampling plans and tables for use in deciding on the acceptability (conformance with specifications) of lots based on the number of defective items observed in specified size of sample(s).

[97] MIL-STD-414, "Sampling Procedures and Tables for Inspection by Variables for Percent Defective," Government Printing Office, Washington, DC 20402.

This publication contains sampling plans and tables for use in decisions of acceptability of lots with respect to nonconformance of items on the basis of the mean and standard deviation of the sample of the measurements.

[98] Mitchell, W.J., R. C. Rhodes, and F. F. McElroy, "Determination of Measurement Data Quality and Establishment of Achievable Goals for Environmental Measurements," In: Quality Assurance for Environmental Measurements, ASTM STP 867, J. K. Taylor and T. W. Stanley, Eds., American Society for Testing and Materials, Philadelphia, PA 19103, pp. 41–52 (1985).

In recognizing the need for measuring and documenting the quality of environmental measurements obtained in its monitoring projects and for setting achievable data quality goals for such projects, the Environmental Protection Agency issues a number of guidance and informational documents addressed to those needs. This paper reviews one of those documents, a comprehensive compilation of the quality of measurements actually obtained with various environmental measurement methods in all environmental media. Included are a brief discussion of the rationale and purpose of the document, limited reviews of the four designated measures of data quality (precision, accuracy, completeness, and method detection limit), and explanations of the format used to list the data quality information. Examples of actual data listings from the document are also included.

[99] Murphy, T.J., "The Role of the Analytical Blank in Accurate Trace Analysis," In: Accuracy in Trace Analysis: Sampling, Sample Handling, and Analysis, NBS Special Publication 422, pp. 509–538, National Bureau of Standards, Gaithersburg, MD 20899 (1976).

This paper discusses the critical importance of reducing the analytical blank and controlling its variability. Practical suggestions are given to control the four major sources of blanks: the environment the analysis is performed in; the reagents used in the analysis; the apparatus used; and the analyst performing the analysis.

[99A] National Bureau of Standards, Special Technical Publication 330, "The International System of Units (SI)" (1937).

This document describes the history of the development of the SI system of units and units outside of the system. The basis for the practical realization of the definitions of some important units is also described.

[100] Natrella, M.G., "Experimental Statistics," NBS Handbook 91, National Bureau of Standards, Gaithersburg, MD 20899.

A practical discussion of statistical evaluation of many kinds of measurement data. Many examples are worked out in detail, and extensive tables are provided for the various statistical tests used in the analysis of data.

[101] NBS, "Brief History of Measurement Systems (With a Chart of the Modernized Metric System)," NBS Special Publication 304A, National Bureau of Standards, Gaithersburg, MD 20899.

A general and easily understandable description of the fundamental system of units and standards for measurement is given.

[102] Nelson, Jr., A.C., D. W. Armentrout, and T. R. Johnson, "Validation of Air Monitoring Data," EPA 600/4-80-030 (1980).

This work provides: a discussion of data validation procedures; selection and implementation procedures; and hypothetical examples and case studies.

[103] Nelson, L.S., "Use of Range to Estimate Variability," *J. Qual. Technol.* 7 (No. 1) (January 1975).

A discussion of the use of the range of a set of measurements to estimate σ. A table of the D_2^* factors needed for this purpose is included.

[104] Nelson, W., "How to Analyze Data with Simple Plots," ASQC Basic References in Quality Control, Statistical Techniques, American Society for Quality Control, Milwaukee, WI 53203.

A nice little book that discusses several graphical ways to look at data when deciding on its distribution.

[105] Oatess, W.E., "Establishment of Accreditation Programs for Environmental Labs," *Environ. Sci. Technol.* 12: 1124–27 (1978).

The cooperation of accreditation organizations, regulatory agencies, consensus standards organizations, and the laboratory community is required to establish credible accreditation programs. The roles each should play are described.

[106] Oppermann, H.K. and J. K. Taylor, "State Weights and Measures Laboratories Program Handbook," NBS Handbook 143, National Bureau of Standards, Gaithersburg, MD 20899 (1985).

Contains criteria for assessing the capability of laboratories (basis for NBS/OWM Certification) and for appraisal for type testing of weights and measures devices.

[107] Paule, R.C. and J. Mandel, "Consensus Values and Weighing Factors," *J. Research NBS*, 87: 377–85 (1982).

A method is presented for the statistical analysis of sets of data which are assembled from multiple experiments. The analysis recognizes the existence of both within-group and between-group variabilities and calculates appropriate weighing factors based on the observed variability for each group. The weighing factors are used to calculate a "best" consensus value from the overall experiment. The technique for obtaining the consensus value is applicable to either the determination of the weighted average value, or to the parameters associated with a weighted least squares regression problem. The calculations are made using an iterative technique with a truncated Taylor series expansion. The calculations are straightforward and are easily programmed on a desktop computer. An examination of the observed variabilities, both within groups and between groups, leads to considerable insight into the overall experiment and greatly aids in the design of future experiments.

[108] Polvi, G.R., Y. S. To, and E. C. Lim, "Overview of the National Discharge Monitoring Report (DMR) Quality Assurance (QA) Program," In: Quality Assurance for Environmental Measurements, ASTM STP

867, J. K. Taylor and T. W. Stanley, Eds., American Society for Testing and Materials, Philadelphia, PA 19103, pp. 189–199 (1985).

As mandated by the Clean Water Act, wastewater treatment facilities are regulated under the National Pollutant Discharge Elimination System (NPDES). Direct discharges have unique NPDES permits with discharge limits and a self-monitoring requirement. The effectiveness of the NPDES Program hinges on the quality of the self-monitoring data. The DMR QA Program evaluates the data quality from the NPDES permittees. Since 1978, the Program has helped to assure the data quality of DMRs. Under this program, all major permittees are annually sent performance samples to be analyzed and reported in the same way as required in the permits. The results of these permittee tests are compared with an established range of acceptance for each test parameter by EPA to identify potential problems with analytical or reporting procedures. EPA/State then follow up through correspondence or on-site visits with permittees to resolve the problems. As an overview, the history, authority, procedures, use of data, and implementation issues are reviewed.

[109] Pontius, P.E. and J. M. Cameron, "Realistic Uncertainties and the Mass Measurement Process," NBS Monograph 103, National Bureau of Standards, Gaithersburg, MD 20899 (1967).

This is the first publication that treats measurement as a process that can be held in a state of statistical control. The concepts involved and the statistical aspects are presented. The concept of statistical measurement control is based largely on the pioneering ideas presented in this paper.

[110] Pontius, P.E., "Notes on the Fundamentals of Measurement as a Production Process," NBSIR 74-545, National Bureau of Standards, Gaithersburg, MD 20899 (1974).

The generalized concept of a measurement process is discussed together with techniques and examples for verifying the validity of the result. While some of the techniques may not be appropriate for certain highly specialized measurement processes, it is felt that the concepts are applicable to virtually all measurement processes which must operate in a variety of environments and must accommodate a variety of materials and properties. The techniques have been invaluable in understanding the manner in which measurement processes operate in the "real" world.

[111] Provost, L.P., "Statistical Methods in Environmental Sampling," In: Environmental Sampling for Hazardous Wastes, ACS Symposium Series 267, American Chemical Society, Washington, DC 20036 (1984).

This paper discusses the role that statistics can play in environmental sampling. Topics considered include: sampling models; determining sample size; allocating sampling resources; strategy for designing sampling plans. The mathematical expressions for computing these and related matters are presented. Because unique solutions are not possible in many cases, nomographs are given that permit quick evaluation of optional approaches in such matters as number of samples, number of measurements, and cost considerations.

[112] Radin, N., "Quality Assurance in Clinical Chemistry—Part I, Workbook Part II, Primer," USHEW, Public Health Service (July 1971).

The primer discusses: measurement variability; the analytical process: statistics, and quality control; and quality assurance. The workbook contains many practical problems to be solved.

[113] Reber, S., "Laboratory Information Management Systems," *Am. Lab.* pp. 78–85 (February 1985).

The general problem of management of data in a large laboratory is discussed, and a computerized system is described that is useful for this purpose.

[114] Rhodes, R.C. and S. Hochheiser, Eds., "Data Validation Conference Proceedings," EPA 600/9-79-042, EPA/EMSL, Research Triangle Park, NC 27711 (September 1979).

The proceedings are a record for future reference of the technical presentations made at a conference on data validation for environmental data. The conference was hosted and sponsored by the U. S. Environmental Protection Agency, Research Triangle Park, Interlaboratory Quality Assurance Coordinating Committee on November 4, 1977, at the Research Triangle Park. Various data validation approaches and techniques were presented and are documented in this publication. A total of 18 papers are included.

[115] Robinson, R.D., D. Knab, and D. R. Perrin, "An Individual Water Sample Quality Assurance Program," Los Alamos National Laboratory, Los Alamos, NM 10163 (March 1985).

This report documents the development and implementation of a flexible Individual Water Sample Quality Assurance (IWSQA) program. The management of the data obtained from the chemical characterization of a water sample is described. These data are used to calculate the IWSQA ratios used to evaluate the data for possible errors and indications of which chemical constituents should be reanalyzed before reporting. The ratios used are: cations/anions; hardness/Ca + Mg; total dissolved solids/defined calculated solids; and conductivity/calculated conductivity. The theory, definitions, tables for calculation, and guidelines for interpretation of the ratios are given.

[116] Rogers, L.B., et al., "Recommendations for Improving the Reliability and Acceptability of Analytical Chemical Data Used for Public Purposes," *Chem. Eng. News* 60 (23): 44 (1982).

Technical problems that relate to identifications and quantitative measurements of low concentrations of chemicals involved in various types of regulations and in clinical samples are addressed in a general way. Recommendations have been developed that should maximize the scientific credibility and general acceptability of the data, as well as their reliability, thereby reducing the controversial aspects of monitoring and/or regulating low concentrations of chemical species. Recommendations emphasize procedures applicable to interlaboratory comparisons.

[117] Sangster, R.C., "Structure and Function of the National Measurement System," NBSIR 75-949, National Bureau of Standards, Gaithersburg MD 20899 (1975).

The general concept of a national measurement system is discussed, i.e., the framework in which compatible measurements can be made.

[118] Schock, H., Ed., "Evaluation and Accreditation of Inspection and Test Activities," ASTM STP 814, ASTM, Philadelphia, PA 19103 (1981).

This is a collection of 21 papers dealing with various aspects of the following subjects: evaluation and accreditation concepts; evaluation and accreditation in government; and international evaluation and accreditation.

[119] Schweitzer, G.E. and J. A. Santolucito, Eds., Environmental Sampling for Hazardous Wastes, ACS Symposium Series 267, American Chemical Society, Washington, DC 20036 (1983).

This 13-chapter book documents the importance of improved approaches to environmental sampling and reviews several successful field programs, including sampling for dioxin in Missouri, lead in Dallas, and cyanide in Washington. Classical and innovative approaches to using statistics in the design and interpretation of monitoring activities are presented. Included as well are well-developed guidelines for applying quality assurance procedures in the field.

[120] Seward, R.W., Ed., "Standard Reference Materials and Meaningful Measurement," NBS SP 408, National Bureau of Standards, Gaithersburg, MD 20899.

This book presents the proceedings of the 6th Materials Research Symposium held at the National Bureau of Standards, Oct. 29–Nov. 2, 1973. It contains 25 papers describing national and international programs for reference material development. The use of statistics, selection criteria, and steps for certifying

SRMs are reviewed. Reports of 15 panel sessions reviewing the use of and needs for reference materials are included.

[121] Shewhart, W.A., Statistical Method from the Viewpoint of Quality Control, The Graduate School, U. S. Department of Agriculture, Washington DC (1939).

This book contains the texts of a series of lectures on the philosophy of statistical product control. It is one of the earliest discussions of this topic that has provided the background for modern quality control.

[122] Shilling, E.G., "A Lot-Sensitive Sampling Plan for Compliance Testing and Acceptance Inspection," *J. Qual. Technol.* 10: 47–51 (1978).

An acceptance sampling plan is presented that is applicable in general acceptance sampling and is particularly useful in compliance and safety-related testing. It provides for: rejection of the lot if any defective items are found in the sample; a well-defined relationship between the sampling plan and the size of the lot being inspected; a clear indication of the economic impact of the quality levels utilized in the plan; and simplicity and clarity in use. Based on the hypergeometric probability distribution, it gives the proportion of the lot that must be sampled to guarantee that the fraction defective in the lot is less than a prescribed limit with odds of 9:1 (i.e., LTPD protection). Various applications are presented, together with a discussion of the operating characteristics and other measures of the plan's protection.

[123] Squires, F.H., Successful Quality Management, Hitchcock Executive Book Service (1980).

A collection of monthly articles appearing in *Quality* magazine. Chapter titles are: Quality Management and You; Executive's Guide to Quality; Product Liability; Employee Motivation; Effective Communication; Inspection and Testing; and Statistics and Sampling. A total of 341 articles are included under the above topics.

[124] NBS, "Standard Reference Materials Catalogue," NBS SP 260, National Bureau of Standards, Gaithersburg, MD 20899.

This is a catalogue describing the SRMs available from NBS.

[125] Stanley, T.W. and S. S. Verner, "The U.S. Environmental Protection Agency's Quality Assurance Program," In: Quality Assurance for Environmental Measurements, ASTM STP 867, J. K. Taylor and T. W. Stanley, Eds., American Society for Testing and Materials, Philadelphia, PA 19103, pp. 12–19 (1985).

The Environmental Protection Agency (EPA) and its predecessor organizations historically have devoted considerable time and effort to the standardization of test procedures, the development and use of calibration and performance audit procedures, quality control and reference samples, training, and other quality assurance (QA) activities. However, these QA activities were essentially voluntary, and most of the data generated were reported with no indication of quality, thereby severely reducing their usefulness. Due to questions concerning the quality of EPA's data, agency policy stipulated on May 30 and June 29, 1979, requires participation in a centrally managed QA program by all organizational units engaged in environmentally related measurements. This policy applies equally to those extramural efforts performed on behalf of EPA. The primary responsibility for program development and direction is assigned to the Office of Research and Development, while other program offices, regions, and laboratories are responsible for its implementation. Essential elements of the program include: developing and implementing QA program plans, QA project plans, and standard operating procedures; conducting audits of the capability and performance of measurement systems, and data quality; maintaining a mechanism for corrective actions; QA training; and frequent reports to management on the quality of data, program effectiveness, and problems. The goal of the QA program is to ensure that all data generated are of known, documented, and acceptable quality.

[126] Taylor, D.R., "Quality Assurance Guidance for the Determination of Inorganic Parameters in Soil, Sediment, and Water Samples," EPA, 600/X-83-013 (February 1983).

The purpose of this document is to describe the QA/QC procedures which should be followed by any laboratory attempting to produce analytical data of known quality from the analysis of inorganic compounds in samples of soil, sediment, water, or hazardous wastes regulated under the Resource Conservation and Recovery Act (RCRA). Most of the QA/QC procedures described herein are those currently used in the National Contracts Analytical Laboratory Program. However, since the NCALP program is undergoing an evolutionary process, the analytical protocols are constantly being improved by input from EPA and contract laboratories, as well as from other sources. The performance control windows for the analytical processes described will be continually updated using a computerized analytical data base compiled at the Environmental Monitoring Systems Laboratory at Las Vegas.

[127] Taylor, J.K., "Validation of Environmental Data by Intercalibration and Laboratory Quality Control Programs," Preprint, ACS Division of Environmental Chemistry, Los Angeles, CA (April 1974).

Accurate methodology, adequate calibrations, and systematic quality control procedures are required to provide reliable analytical data. These three factors may be considered as links in a chain that is only as strong as the weakest segment. Well-established measurement processes are conventionally established by deliberate or evolutionary processes, but environmental analysis has weak foundations in each of the three areas. The rapid growth of this field, both in the number of analyses required and in the variety of substances that must be controlled or measured to evaluate their actual or potential environmental impact, is largely responsible for this situation. Legislative requirements have often forced the acquisition of data that would have better remained unreported. The situation needs to be rectified by a systems approach that assures the validity of the data reported by individual stations and permits its incorporation into the body of data derived from diverse sources. This paper discusses the measurement system with the objective of encouraging improvements where needed.

[128] Taylor, J.K., "Importance of Intercalibration in Marine Analysis," *Thal. Jugo.* 14: 221–29 (1978).

This paper discusses the general principles of intercalibration, the process by which measurement stations may be calibrated with respect to each other. When intercalibrated, the measurements generated may be considered as mutually compatible and admissible into a body of data for compilation or other purposes. The process involves the use of recognized sampling techniques and measurement methods, as well as rigorous quality control procedures within each measurement station. Standard reference materials can provide the intercalibration linkage to verify that compatibility of data does exist.

[129] Taylor, J.K., "Quality Assurance of Chemical Measurements," *Anal. Chem.* 53: 1588A-96A (1981).

This is a general article describing the basic aspects of quality assurance as related to chemical analysis.

[130] Taylor, J.K., "Quality Assurance Measures for Environmental Data," In: Lead in the Marine Environment, M. Branica and Z. Konrad, Eds., Pergamon Press, pp. 1–7 (1980).

An understanding of the occurrence, fate, effects, and pollution potential of suspected toxic substances requires a wide variety and large quantity of analytical data obtained over extended time intervals, in diverse locations, and by independent investigators. Unfortunately, much of the data reported in the literature is often difficult to interpret and correlate because of lack of information on the reliability of sampling and measurement. This paper discusses general principles of quality control of both measurement and sampling, and recommends minimum requirements for reporting environmental data. The use of generally accepted reference materials to intercalibrate measurement laboratories is emphasized.

[131] Taylor, J.K., "Reference Materials—What They Are and How They Should be Used," *J. Testing Eval.* 11: 355–7 (1983).

Reference materials consist of known samples that may be used to verify the accuracy of measurements of a system known to be in statistical control. They are also useful for evaluating methodology and for evaluating the proficiency of analysts and laboratories. Ways to use reference materials systematically are described.

[132] Taylor, J.K., "Principles of Quality Assurance of Chemical Measurements," NBSIR 85-3105 National Bureau of Standards, Gaithersburg, MD 20899.

The general principles of quality assurance of chemical measurements are discussed. They may be classified as quality control—what is done to control the quality of measurement process, and quality assessment—what is done to evaluate the quality of the data output. Quality assurance practices are considered as a hierarchy with levels progressing from the analyst, to the laboratory, to the project, to the program. The activities of each level are different and depend upon the ones beneath it. Recommendations are presented for developing credible quality assurance practices at each level. An appendix contains outlines that may be used to develop the various documents associated with a quality assurance program. An extensive list of definitions of terms used in quality assurance is included.

[133] Taylor, J.K., "What is Quality Assurance?" In: Quality Assurance for Environmental Measurements, ASTM STP 867, J. K. Taylor and T. W. Stanley, Eds., ASTM, Philadelphia, PA 19103, pp. 5-11 (1985).

The quality of data must be known and established before it can be used logically in any application. Data quality may be judged on the basis of its quantitative accuracy and on the confidence that can be placed in the qualitative identification of the parameters measured. This requires the production of data in a quality assurance program that permits the assignment of its statistically supported limits of uncertainty. The essential features of such a program, consisting of quality control and quality assessment techniques, are discussed in the paper.

[134] Taylor, J.K., "Handbook for SRM Users," NBS SP 260-100, National Bureau of Standards, Gaithersburg, MD 20899 (1985).

This handbook gives a comprehensive discussion of the use of reference materials to assess the accuracy of a measurement system. Also included are the selection of SRMs, the certification process, and the measurement of certified values. An appendix reviews the statistical procedures and statistical tables useful in evaluating results of measurements.

[135] Taylor, J.K., "Validation of Analytical Methods," *Anal. Chem.* 55: 600A-608A (1983).

Validation is defined as the process of ensuring that methodology is potentially useful for a given purpose. The hierarchy of analytical methodology is discussed as technique, method, procedure, and protocol, each of which should be validated.

[136] Taylor, J.K., "Guidelines for Evaluating the Blank Correction," *J. Testing Eval.* 12: 54–5 (1984).

The statistical considerations in applying the blank correction in trace analysis are discussed. The question of acceptable limits for the blank is addressed. Unless sufficient measurements are made, the uncertainties in the blank correction may be the major source of uncertainty in ultra-trace analysis.

[137] Taylor, J.K., "Essential Features of a Laboratory Quality Assurance Program," In: Statistics in the Environmental Sciences, ASTM STP 845, S. M. Gertz and M. D. London, Eds., ASTM, Philadelphia, PA 19103.

Progress in the environmental sciences is vitally dependent on reliable data resulting from complex measurement processes. Because of this complexity, the measurement process must be well designed and operate in a state of statistical control. A quality assurance program, including quality control and quality assessment procedures, denotes those features that lead to the production of data under these conditions. The rudimentary features are described together with the expected benefits. Parallelisms are drawn with a well-designed manufacturing process.

[138] Taylor, J.K., "Quality Assurance for a Measurement Program," In: Environmental Sampling for Hazardous Wastes, G. E. Schweitzer and J. A. Santolucito, Eds., ACS Symposium Series 267, American Chemical Society, Washington, DC 20036 (1985).

Quality assurance for measurement programs requires participants of demonstrated competence using appropriate quality assurance procedures. Responsibilities of participants and managers are discussed. The use of a reference laboratory to assure that program objectives are met is described and recommended.

[139] Taylor, J.K., "Evaluation of Data Obtained by Reference Methods," In: Calibration in Air Monitoring, ASTM STP 598, ASTM, Philadelphia, PA 19103 (1976).

The key role of reference methods for measurement of air pollutants in regulatory matters demands that the data obtained in their use be precise and accurate, but no procedures have been established for evaluating their reliability and validity for the intended use. This paper discusses the general principles of reliable analytical measurements and presents guidelines by which the quality of data obtained by reference methods or other procedures may be evaluated.

[140] Taylor, J.K. and T. W. Stanley, Eds., "Quality Assurance for Environmental Measurements," STP 867, American Society for Testing Area Materials, Philadelphia, PA 19103 (1985).

This book contains the text of 36 papers presented at a conference of the same title at the University of Colorado, Boulder, CO, in August 1983. The contents are as follows: General Topics – 2 papers; Data Quality Assessment – 3 papers; Ambient Air Measurements – 3 papers; Ambient Water Measurements – 4 papers; Source Measurements – 4 papers; Discharge Monitoring – 6 papers; Quality Assurance Management – 2 papers.

[141] Taylor, J.K., The Quest for Quality Assurance, American Laboratory, pp. 67–75 (October 1985).

The basic need for a measurement assurance program is discussed, and various approaches are described. The ethical and professional responsibilities of analytical chemists are discussed, and a code of ethics for the practice of analytical chemistry is presented.

[142] Taylor, J.K., "The Role of Collaborative and Cooperative Studies in Evaluation of Analytical Methods," *J. Assoc. Off. Anal. Chem.* 69: 398 (1986).

A method proposed as a standard or for use in a regulatory process or other purposes must be reliable, and its typical performance characteristics should be stated and verified. Collaborative testing is the most acceptable way to accomplish the latter, but its function should not be misunderstood. Such testing can verify performance characteristics and experimentally demonstrate that the methodology can be used successfully by a representative group of laboratories. It does not necessarily support the validity of any data obtained using the method because this may depend on many other factors, including the expertise of the laboratory and the quality assurance aspects of its measurement process.

[142A] Taylor, J.K., "The Critical Role of Samples to Environmental Decisions," *TRAC* 8 (No.5): 121–3 (May 1986).

Samples may be collected according to three kinds of plans: intuitive; statistical; and protocol. Only statistical samples can be interpreted statistically. All others must be interpreted on the basis of judgment which is open to differences of opinion. Often, statistical plans require more samples than feasible, so hybrid plans may be used. The consequences of this must be considered when interpreting the results of such measurements.

[143] Tietz, N.W., "A Model Comprehensive Measurement System in Clinical Chemistry," *Clin. Chem.* 25/6: 833–39 (1979).

A conceptual system of employing validated methods and reference materials is discussed to provide the accuracy required of clinical measurements. Topics discussed include definitive methods and reference materials. A measurement system for the determination of calcium in body fluids, based upon the above principles, is described.

[144] Visman, J., "A General Sampling Theory," *Mat. Res. Stds.* p. 8 (November 1969).

A general theory is presented for sampling based on the premise that the variance in a bulk lot of a material can be expressed by a relation incorporating two sampling constants. The relation may be used to design sampling programs and to determine in advance the precision of a sampling experiment as a function of sample size and the number of increments. See also discussions of this in the articles listed below:
Duncan, A. J., Comments on "A general sampling theory," *Mat. Res. Stds.* 11(1):25 (1971).
In this note, it is shown that the empirical results of Visman can in part be derived from statistical theory.
Visman, J., A. J. Duncan, and M. Lerner, Further discussion: "A general theory of sampling," *Mat. Res. Stds.* 11(8):32 (1971).
In this note, Visman and Duncan discuss the generality of the Visman approach as well as its validity under certain circumstances. Lerner summarizes their points of view and concludes that the Visman procedure, while not strictly general, is a giant step above the previous rules-of-thumb often used.
Visman, J., and A. J. Duncan, Discussion 3 on "A general theory of sampling," *J. Mat.* 7:345 (1971).
Here, Visman develops further the factors that hold the constant relating to segregation variance at a stable value. Duncan agrees with Visman's treatment and conclusions, and develops the statistical theory for Visman's approach.

[145] Wening, R.J., "The Role of the Quality Control Manual in the Inspection and Testing Laboratory," In: NBS SP 591, pp. 99–103. (*See* reference 24.)

The QA manual is the cornerstone of a laboratory's credibility and an important ingredient in its assessment for accreditation. Its content and mode of use are described.

[146] Wernimont, G., "Ruggedness Evaluation of Test Procedures," *ASTM Standardization News*, pp. 13–16 (March 1977).

A discussion of Youden's ruggedness test with extension to testing 18 variables.

[147] Western Electric, Statistical Quality Control Handbook—Select Code 700-444, Western Electric Co., Commercial Sales Clerk, P.O. Box 26205, Indianapolis, IN 46226.

This book contains a wealth of information on the applications of control charts in monitoring industrial production processes. Particularly useful is the extensive treatment of diagnosis of production problems via control charts. While written for engineers operating industrial processes, the extension of the ideas to measurement processes is obvious in most cases.

[148] Westgard, J.D. and T. Groth, "A Multi-Rule Shewhart Chart for Quality Control in Clinical Chemistry," *Clin. Chem.* 27: 493–501 (1981).

This work presents a decision chart and suggests several courses of action of use when interpreting control chart data.

[149] Williams, L.R., "Guidelines for Conducting Collaborative Testing of Biological Test Methods," EPA 600/X-83-029 (1983).

This summarizes the approach, approved by EPA's Office of Toxic Substances, to use when conducting collaborative tests of biological methods.

[150] Williams, L.R., "Quality Assurance Considerations in Conducting the Ames Test," In: Quality Assurance for Environmental Measurements, ASTM STP 867, J. K. Taylor and T. W. Stanley, Eds., American Society for Testing and Materials, Philadelphia, PA 19103, pp. 260–271 (1985).

This paper provides general and specific guidance for the development and implementation of a quality assurance program for Ames testing. An effective quality assurance program is needed to assure that data produced from short-term mutagenicity testing (e.g., the Ames test) are of known quality and comparable among laboratories conducting the testing. The success of such a program begins with a personal and organizational commitment to data quality. Next, standardized procedures for the conduct and control of the testing process should be adopted and rigidly followed. Finally, participation in an external quality assurance program is necessary for independent confirmation of testing performance and confirmation of interlaboratory data comparability. Comparability of test data is required if there is to be confidence in the results of testing activities and acceptance of the resulting interpretation of test findings. The functions and responsibilities of an effective quality program are outlined. Primary functions include development of standards based upon technically attainable criteria, assistance to participating laboratories in adopting those standards, and independent evaluation of the success of the laboratories in meeting those standards.

[151] Wilson, A.L., "Approach for Achieving Comparable Analytical Results from a Number of Laboratories," *Analyst* 104: 273 (1979).

QA in a water lab is discussed. Close attention to many details is stressed.

[152] Wilson, A.L., "The Performance Characteristics of Analytical Methods, I." *Talanta* 17: 21–29 (1970).

The ever increasing volume of analytical literature makes it important to be able to compare unambiguously the advantages and disadvantages of analytical methods. To this end, a set of consistent definitions and methods for determining quantitative performance characteristics (e.g., precision, sensitivity, bias) is needed. The aim of this series of papers is to review the definition and determination of such parameters and to suggest criteria for general use. This first paper discusses the general problem, considers those general aspects of analytical methods that are important, and establishes the performance characteristics to be considered in detail.

[153] Wilson, A.L., "The Performance Characteristics of Analytical Methods, II" *Talanta* 17: 31–44 (1970).

Statements on the errors of analytical results are an important aspect of characterizing the performance of analytical methods. The general nature of random and systematic errors is briefly discussed, and methods of numerically defining the former are considered. It is suggested that the standard deviation of analytical results be used exclusively as the quantitative measure of precision within the context of performance characteristics. Techniques for, and precautions to be observed in, estimating standard deviation are critically discussed. On this basis, general principles are proposed that should be observed whenever possible in experimental tests to estimate standard deviation.

[154] Wilson, A.L., "The Performance Characteristics of Analytical Methods, III" *Talanta* 20: 725–32 (1971).

The range of concentrations covered by an analytical method is an important performance characteristic, and suggestions are made for definition of this range. The ability to detect small concentrations is also considered, and it is also suggested that the standard deviation of blank determinations be quoted as the performance characteristic relevant to the power of detection.

[155] Youden, W.J., "Collection of Papers on Statistical Treatment of Data," *J. Qual. Technol.* 4 (No. 1): 1–67 (January 1982).

This memorial issue of the Journal of Quality Technology reviews the contributions of the late W. J. Youden to the statistical treatment of measurement data. It contains a biography, selected reprints, and a complete bibliography of his publications.

[156] Youden, W.J., "Experimentation and Measurement," NBS SP 672, National Bureau of Standards, Gaithersburg, MD 20899 (1984). Reprint of a book originally published by the National Science Teachers Association.

A very readable primer written for the high school student to introduce statistical ideas with practical examples.

[157] Youden, W.J., "Ranking Laboratories by Round Robin Tests," *Mat. Res. Stds.* (January 1963).

A procedure is described to evaluate the performance of laboratories by ranking them according to the magnitude of the results they report on a series of test samples. A statistical test to recognize laboratories that consistently report low and high results is given.

[158] Youden, W.J., "Statistical Techniques for Collaborative Tests," AOAC, Arlington, VA 22209.

The philosophy of collaborative testing is discussed together with the statistical evaluation of test data. This is the predecessor of the publication in Reference 159, which also should be consulted in this regard.

[159] Youden, W.J. and E. H. Steiner, "Statistical Manual of the AOAC," AOAC, Arlington, VA 22209 (1975).

This is an extension of the first edition (by Youden) which discusses the philosophy of collaborative testing and presents AOAC guidelines for planning and conducting a collaborative test.

[160] Zacherle, A.W., "A Summary of Federal Quality Assurance Programs Related to Marine Pollution," NOAA, National Marine Pollution Program, 11400 Rockville Pike, Rockville, MD 20852.

This report describes the quality assurance practices being followed by seven U. S. government agencies that are engaged in various measurement activities related to marine pollution research.

INDEX*

*Numbers in italic refer to references in Appendix E, page 289